MATHEMATICS

BOOSTER-1

MANOJ KUMAR SINGH

ADHYYAN BOOKS

Mathematics Booster-1

1st Edition

Publication Date: August 2018

Price: ₹500 | $14.99

ISBN: 978-93-87502-60-4

Published by:

Adhyyan Books

Office No. 637,

Opposite Vivanta by Taj,

DDA SFS. Pocket-1, Dwarka,

Sec-22, New Delhi-110077

Website: www.adhyyanbooks.com

E-mail: contact@adhyyanbooks.com

Dedication

This book **"Mathematics Booster-1"** has been written for enhancing calculations and creating affection for mathematics.

Fear of mathematics is a mere confusion we develop. Instead, it is amusing and develops overall analytical capability of human mind.

This book is dedicated to Author's Parent and fellow citizens of village **"Bishunpura"**.

Author Desk

Being a Mechanical Engineer with 27th Ranked in BCECE-2K4 and graduated from BIT Sindri – Dhanbad, Entrepreneurship program from Symbiosis, Pune. My professional career continued with leading Indian corporates.

Apart from these, my passion thrived me for being *Life Coach*, an *Author* and a *blogger*. Conceptualized "PARV-Quantifying learning ability" and Contributed to develop www.24sai.com (Strengthening Academics and Institutions – Online learning and teaching tool).

Many competitive exam aspirants, students, Kids struggle with calculation, this book will help students to learn mechanics behind mathematics.

Appreciating effort of Sanika Sonawane (onc of my student, who has been winner at various mathematical competitions) for contributing in content review, editing and sharing Insights.

Manoj Singh

Contents

Introduction

Mathematics is one of the most important subject in our life. No matter who we're and wherein career we're operating. Buying grocery for kitchen or a massive billion enterprise deal, all contain mathematics as a primary requirement for our daily or lifelong desires.

From early childhood stage to entire span of our life, we in most cases engage with calculations even though we have exclusive professions and livelihood. Any of us hardly remember who taught us first - The digits 1,2,3,4,5-------. When you grow in family, it be most lovely thing to see you counting numbers and letters with mumbling sounds and ignoring the questions asked multiple times by elders. As a kid, most of us try to escape from these activities of counting numbers and chanting letters however our elders (Parent, teacher, siblings, relatives, neighbors....) continue the desire of coaching us counting in our early stage.

For most of us, the adventure of learning mathematics starts from the time while we will in no way recollect, how the things were accomplished and taught in our childhood.

All of us had started out counting numbers on finger or counting some objects, chocolates, sweets and so on.... We all start off evolved calculating in similar way however few of us turns into master in mathematics whilst few contains amazing fear for this.

Can we imagine, how does it occur (Enjoying Calculation for few while fear for others)?

Well, let's recognize the reality at the back of it. Mathematical calculations may be carried out in many ways. Few of methods are hard, few easy and simple to understand at the same time some miss-concepts always lead us to wrong method or result.

As we grow older, most of us learn wrong-concepts or tough methods of solving problems while we need to understand all different methods to enhance our concepts, learnings and calculations.

Here we will practice different methods for basic mathematics. This could assist each one of us for quick calculations. Competitive exams have time limits to solve the questions. Each second helps aspirants to steer over different competitors. Quick calculations save student's time. It also helps to analyze even a list of long data or difficult problems. It creates confidence, fun and capabilities for mathematics.

We all had begun counting1, 2, 3, 4, 5, 6.......... in our childhood. Further we all learnt 0 (Zero) as well but can you guess whether these numbers 0,1,2,3,4,5,6,7,8,9....... are mere counting table or **addition table.**

if we guess efficaciously then we come to recognize that these table were addition table, 0+1 = 1, 1+1=2, 2+1=3, 2+3 = 5, 8+9=17, 20+80=100 etc.. We learn the counting table in such a way that addition of numbers turns into most smooth step in mathematical calculations. Even if we need to add higher numbers, we experience it easy and simple. We must thank all our teachers, family and relatives who taught us addition table and created confidence of solving additions of numbers.

We grew up solving addition of numbers like 98+57=155 in an easy manner as we do 8+9=17 or 15+17=32.

In childhood, we learnt unit, tens, hundreds, thousands.... places for any number. For example:

In number like 9876

Number at unit place is = 6

Number at tens place is = 7

Number at hundreds place is = 8

Number at thousands place is = 9

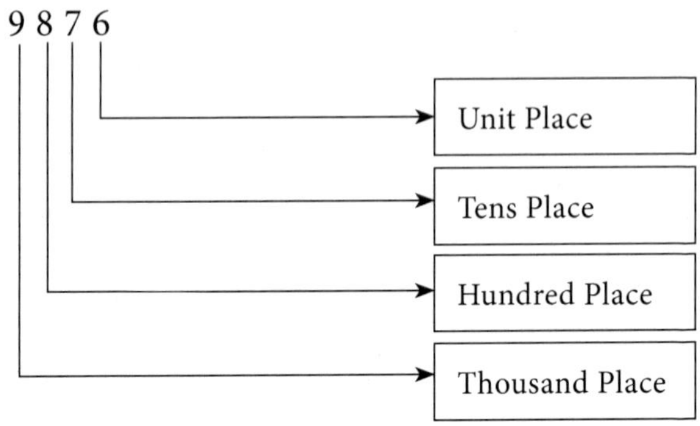

So, we go from unit place to higher ones (tens, hundreds, thousands...) from right to left always. We follow the same sequence for bigger numbers like ten thousand, lakh, tens lakh, crore, ten crores etc.....

Representation of numbers by expanding in their respective place values has been shown below.

9876 = 9x1000 + 8x100 + 7x10 + 6

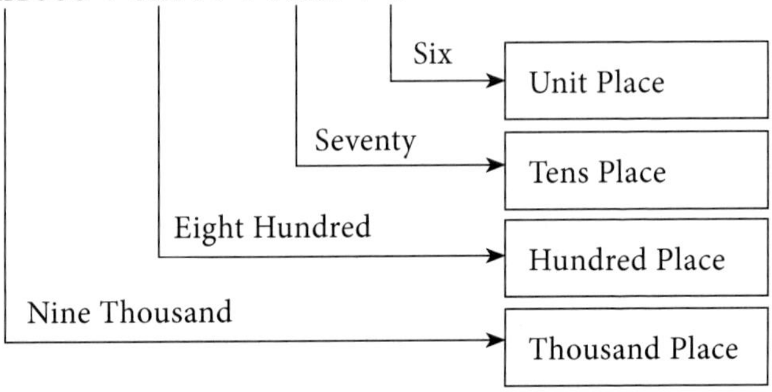

And this is read as "Nine thousand eight hundred and seventy-six".

Here is a small exercise for you to try and examine your skills of understanding the concept of place values.

?	**You may try expanding following numbers** **a. 7864 b. 976 c. 298746 d. 956429** **e. 35736**

CHAPTER 1

Representation of Numbers

Indian Format

How do we put up commas in Indian mathematical format?

We represent numbers with commas at specified places in such a way that there is a comma after first three digits from right and again a comma after the next two

Thus, we can also read and write numbers as shown below:

9876 can be written as 9,876 and read as nine thousand eight hundred and seventy-six.

79876 can be written as 79,876 and read as seventy nine thousand eight hundred and seventy six.

579876 can be written as 5,79,876 and read as five lakh seventy nine thousand eight hundred and seventy six

So the above numbers can be written as

9876 = 9,876 = 9x1000 + 8x100 + 7x10 + 6

and read as Nine thousand eight hundred and seventy six

79876= 79,876= **79 x 1000 + 8x100 + 7x10 + 6**
= 7x10000 + 9x1000 + 8x100 + 7x10 + 6

and read as Seventy nine thousand eight hundred and seventy six

579876=5,79,876= **5x100000 + 79 x 1000 + 8x100 + 7x10 + 6** = 5x100000 + 7x10000 + 9x1000 + 8x100 + 7x10 + 6

Read as Five lakh seventy nine thousand eight hundred and seventy six

Similarly we can represent higher numbers as

6579876= 65,79,876 = **65x100000 +79 x 1000 + 8x100 + 7x10 + 6**

and read as Sixty five lakh seventy nine thousand eight hundred and seventy six

46579876= 4,65,79,876= **4x10000000 + 65x100000 + 79 x 1000 + 8x100 + 7x10 + 6**

and read as Four crore sixty five lakh seventy nine thousand eight hundred and seventy six

246579876= 24,65,79,876= **24x10000000 + 65x100000 + 79 x 1000 + 8x100 + 7x10 + 6** and read as Twenty four crore sixty five lakh seventy nine thousand eight hundred and seventy six

?	**Put commas at appropriate place and write the number in words.** **a. 7864676** **b. 9767697** **c. 298746** **d. 95647698** **e. 3573689** **f. 78677** **g. 867969**

In Indian mathematical format, we put commas at thousands place, lakhs place, crores place and so on as discussed above.

So, commas are represented as

1,000 = One thousand

10,000 = Ten thousand

1,00,000 = One lakh

10,00,000 = Ten lakh

1,00,00,000 = One crore

10,00,00,000 = Ten crore

?	**Try following numbers to write with commas, expand and write in words in Indian format:** **a. 7698753 =** **b. 876543297 =** **c. 365789 =** **d. 10657 =** **e. 753124 =**

British/American Format

How do we put up commas in British / American format?

In Indian mathematical format, we have learnt putting first comma after three digit from right and second comma after next two digit, third comma after further next two digit and so on.

But in British / American system, we put up commas after every three digits.

Indian Number representation Vs British / American number representation

Indian number representation Vs British / American Number in short scale

Indian	***Notation with comma***	***British/ American***	***Notation with comma***
Ten	10	Ten	10
Hundred	100	Hundred	100
Thousand	1,000	Thousand	1,000
Ten thousand	10,000	Ten thousand	10,000
Lakh	1,00,000	Hundred Thousand	100,000
Ten Lakh	10,00,000	Million	1,000,000
Crore	1,00,00,000	Ten Million	10,000,000
Ten crore	10,00,00,000	Hundred Million	100,000,000
Hundred crore	1,00,00,00,000	Thousand million or One Billion	1,000,000,000

We can write following numbers in British / American format of short scale

6579876= 6,579,876 = **6x1000000** + **579 x 1000 +**
8x100 + 7x10 + 6

and read as Six million five hundred seventy-nine thousand eight hundred seventy-six

46579876= 46,579,876= **46x1000000 + 579 x 1000 +**
8x100 + 7x10 + 6

and read as Fourty six million five hundred seventy-nine thousand eight hundred seventy-six

246579876= 246,579,876= 2**46x1000000 +**
579 x 1000 + 8x100 + 7x10 + 6

and read as Two hundred fourty-six million five hundred seventy-nine thousand eight hundred seventy-six

Number	**Indian System**	**British / American system-Short scale**
235	235 = Two hundred and thirty five	235 = Two hudred-thirty five
4235	4,235 = Four thousand two hundred and thirty five	4,235 = Four thousand two hundred thirty -five
64235	64,235 = Sixty four thousand two hundred and thirty five	64,235 = Sixty four thousand two hundred thirty -five

Number	**Indian System**	**British / American system-Short scale**
864235	8,64,235 = Eight lakh sixty four thousand two hundred and thirty five	864,235 = Eight hundred Sixty -four thousand two hundred thirty -five
7864235	78,64,235 = Seventy eight lakh sixty four thousand two hundred and thirty five	7,864,235 = Seven million Eight hundred Sixty -four thousand two hundred thirty -five
27864235	2,78,64,235 = Two crore seventy eight lakh sixty four thousand two hundred and thirty five	27,864,235 = Twenty seven million Eight hundred Sixty -four thousand two hundred thirty -five
527864235	52,78,64,235 = Fifty two crore seventy eight lakh sixty four thousand two hundred and thirty five	527,864,235 = Five hundred twenty -seven million eight hundred sixty -four thousand two hundred thirty -five
8527864235	8,52,78,64,235 = Eight arab fifty two crore seventy eight lakh sixty four thousand two hundred and thirty five	8,527,864,235 = Eight billion five hundred twenty -seven million eight hundred sixty -four thousand two hundred thirty -five

Number	**Indian System**	**British / American system-Short scale**
68527864235	68,52,78,64,235 = Sixty eight arab fifty two crore seventy eight lakh sixty four thousand two hundred and thirty five	68,527,864,235 = sixty eight billion five hundred twenty -seven million eight hundred sixty -four thousand two hundred thirty -five
568527864235	5,68,52,78,64,235 = Five kharab sixty eight arab fifty two crore seventy eight lakh sixty four thousand two hundred and thirty five	568,527,864,235 = Five hudred sixty -eight billion five hundred twenty -seven million eight hundred sixty -four thousand two hundred thirty -five
9568527864235	95,68,52,78,64,235 = Ninety five kharab sixty eight arab fifty two crore seventy eight lakh sixty four thousand two hundred and thirty five	9,568,527,864,235 = Nine Trillion five hudred sixty -eight billion five hundred twenty -seven million eight hundred sixty -four thousand two hundred thirty -five

CHAPTER 2
Addition of Numbers

After reading and writing numbers in different formats as shown above, we will now learn to add numbers.

Is addition of Numbers tough?

Most of us will reply "No" and will say that It's easy to sum up numbers of any order.

In childhood days, we learnt addition or summation of numbers perfectly.

Then why do I need to discuss methods of addition here?

When a student appears in an academic examination or any competition, He or she needs to calculate fast. A quick method of addition helps in saving time as well as buildup confidence in students.

Sometimes, it's also fun and enjoyment when we solve and add some numbers much faster than others.

How do we add numbers and what are other different methods for quick calculations?

Here we will discuss five different methods for addition:

1. **Addition with Carry over method**
2. **Addition without carry over method**
3. **Addition with currency count method**
4. **Digit sum method**
5. **Addition of Number by converting Higher digit to lower digit**

Addition with Carry Over Method

Almost all students learn the first method as addition with Carry over.

When we need to add 536 + 987 then we follow these steps.

We add unit number first from right side in both given numbers

(Unit number in 536 is 6 , and unit number in 987 is 7, we add 6 +7 = 13 and write 3 at units place with a carry over 1 for tens place .

Step-1

$$\begin{array}{r} 5\ 3\ 6 \\ +\ 9\ 8\ 7 \\ \hline \end{array}$$

Step-2 (Addition of numbers at unit place, 6+7 = 13, put 3 at unit place and carry 1 at tens place.

$$\begin{array}{cccc} & & & 1 \\ & 5 & 3 & 6 \\ + & 9 & 8 & 7 \\ \hline & & & 3 \end{array}$$

Step-3 (Addition of numbers at tens place, 1+3+8 = 12, put 2 at tens place and carry 1 at hundred place.

$$\begin{array}{cccc} & & & 1 \\ & 5 & 3 & 6 \\ + & 9 & 8 & 7 \\ \hline & & 2 & 3 \end{array}$$

Step-4 (Addition of numbers at hundred place , 1+5+9 = 15 , now put 15 at hundred place as no further numbers available for addition.

$$\begin{array}{cccc} & & & 1 \\ & 5 & 3 & 6 \\ + & 9 & 8 & 7 \\ \hline 1 & 5 & 2 & 3 \end{array}$$

So the answer for 536 + 987 = 1523.

We all have learn this method in our school days.

Addition without Carry Over Method

Few students face difficulties in carry over method.

This method may help them learn addition in a much better way.

Again, we consider the same example of adding 536 + 987

This is a two stage addition process, we simply add the numbers at units place and write them down below the number as a whole at units place, similarly we add tens numbers and write them down below ending at tens place and so on...

Step-1

```
    5 3 6
  + 9 8 7
  -------
```

Step-2 (Addition of numbers at units place, 6+7 = 13, put 13 at units place in result)

```
    5 3 6
  + 9 8 7
  -------
      1 3
```

Step-3 (Addition of numbers at tens place, 3+8 = 11, put 11 at tens place in result)

```
    5 3 6
  + 9 8 7
  -------
      1 3
    1 1
```

Step-4 (Addition of numbers at hundreds place, 5+9 = 14, now put 14 at hundreds place)

```
    5 3 6
  + 9 8 7
  -------
      1 3
    1 1
  1 4
```

Step-5 (Now add numbers below to get the answer)

```
    5  3  6
 +  9  8  7
 ----------
       1  3
    1  1
 1  4
 ----------
 1  5  2  3
```

So the answer for 536 + 987 = 1523.

In this method, student can start adding numbers from any place value numbers they choose like units, tens, hundred place ...

For example

Step-1

```
    5  3  6
 +  9  8  7
 ----------
```

Step-2 (Addition of numbers at hundreds place, 9+5 = 14, put 14 at hundreds place)

```
    5  3  6
 +  9  8  7
 ----------
 1  4
```

Step-3 (Addition of numbers at tens place, 3+8 = 11 , put 11 at tens place)

```
    5  3  6
 +  9  8  7
 ----------
 1  4
    1  1
```

Step-4 (Addition of numbers at units place, 7+6 = 13, now put 13 at hundred place)

$$\begin{array}{cccc} & 5 & 3 & 6 \\ + & 9 & 8 & 7 \\ \hline 1 & 4 & & \\ & 1 & 1 & \\ & & 1 & 3 \\ \hline \end{array}$$

Step-5 (Now add numbers below to get the answer)

$$\begin{array}{cccc} & 5 & 3 & 6 \\ + & 9 & 8 & 7 \\ \hline 1 & 4 & & \\ & 1 & 1 & \\ & & 1 & 3 \\ \hline 1 & 5 & 2 & 3 \end{array}$$

isn't it fun to just start adding numbers without thinking about which place value it belongs?

? **Add following number without carry over method.**

a. 7864 + 1135 b. 2376 + 5423 c. 2653 + 8246

d. 956+ 132 e. 357+ 342 f. 78677 + 11221

Addition with Currency Count Method

(This is also method for addition from left to right)

This method may help students to convert the sum into nearest currency value or nearest multiple of 10, 100,

1000... (Eg. 10 , 20, 30 , 40 , 50 , 60 , 70 ,80 , 100......1000... and so on)

Again we may consider the same example to add 536 + 987

We can convert 536 as

$$536 = 500 + 30 + 6$$

And 987 can be converted as

$$987 = 900 + 80 + 7$$

So we can add 536 + 987 as

$$\begin{aligned} 536 + 987 &= 500 + 30 + 6 + 900 + 80 + 7 \\ &= 900 + 500 + 80 + 30 + 7 + 6 \\ &= 1400 + 110 + 13 \\ &= 1400 + 100 + 10 + 10 + 3 \\ &= 1500 + 20 + 3 \\ &= 1500 + 23 \\ &= 1523 \end{aligned}$$

this is also a method of addition from left to right as shown above. We use this method knowingly or unknowingly for calculating currency in our day to day life.

Step-1

$$\begin{array}{r} 5\ 3\ 6 \\ +\ 9\ 8\ 7 \\ \hline \end{array}$$

Step-2 (Addition of numbers at hundred place 500 + 900 = 1400, put 1400 ending zero at **units place**)

$$\begin{array}{cccc} & 5 & 3 & 6 \\ + & 9 & 8 & 7 \\ \hline 1 & 4 & 0 & 0 \end{array}$$

Step-3 (Addition of numbers at tens place, 30+80 = 110, put 110 at **unit place**)

$$\begin{array}{cccc} & 5 & 3 & 6 \\ + & 9 & 8 & 7 \\ \hline 1 & 4 & 0 & 0 \\ & 1 & 1 & 0 \end{array}$$

Step-4 (Addition of numbers at unit place, 7+6 = 13, now put 13 at **unit place**)

$$\begin{array}{cccc} & 5 & 3 & 6 \\ + & 9 & 8 & 7 \\ \hline 1 & 4 & 0 & 0 \\ & 1 & 1 & 0 \\ & & 1 & 3 \\ \hline \end{array}$$

Step-5 (Now add number below to get the answer)

$$\begin{array}{cccc} & 5 & 3 & 6 \\ + & 9 & 8 & 7 \\ \hline 1 & 5 & 2 & 3 \end{array}$$

?	**Add following number with currency count method.**
	a. 535 + 275 b. 195 + 105 c. 2565 + 3434
	d. 956+ 132 e. 357+ 342 f. 78677 + 11221

Digit Sum Method

This is the most interesting method for addition but it should be used only to verify answers. This is not helpful in calculations but it surely saves time in multiple choice questions to select correct answer.

To begin with this method let's take an example:

Which is correct sum for 536 and 987?

1. 1523
2. 3425
3. 4623
4. 5326

We may simply add and get the answer as this question involves addition of two numbers only.

What if we need to find correct answer for a sum of many numbers.

For example

Which option is correct for the sum 54326 + 32696 + 92765 + 85678 + 72567 + 92356 ?

1. 430388
2. 250438
3. 324648
4. 235678

Here you may try solving the above sum and check the time consumed in solving the problem.

Let us discuss "Digit Sum" Method.

Digit sum of any number = sum of digits of the number.

In digit sum we consider the numbers as follows:

Numbers	*Digit Sum*
0	0
1	1
2	2
3	3
4	4
5	5
6	6
7	7
8	8
9	0

if we try to solve the above question then the digit sum goes as follows:

Digit sum of 54326 = 5+4+3+2+6 = 20=2+0=2

Digit sum of 32696 = 3+2+6+9+6=26=2+6=8

Digit Sum of 92765 = 0+2+7+6+5=20=2+0=2 (we count 0 for 9)

Digit sum of 85678 = 8+5+6+7+8 = 34=3+4=7

Digit Sum of 72567 = **7+2**+5+6+7=0+5+6+7=18=1+8=9=0

Digit Sum of 92356 = 0+2+3+5+6=16=1+6=7

For any addition to be correct.

Digit sum of the left-hand side (I.e. the question) should be equal to the right-hand side (I.e. the answer)

Digit sum in left = Digit Sum in right.

Therefore, digit sum for

54326 + 32696 + 92765 + 85678 + 72567 + 92356

=2+8+**2+7**+0+7

=**2**+8+0+0+**7**

=0+8+0+0

=8

Now if we check the options then

Digit sum for option 1 = 4+3+0+3+8+8= 26=2+6=8

Digit sum for option 2 = 2+5+0+4+3+8 =20=2+0=2

Digit sum for option 3 = 3+2+4+6+4+8 = 27=2+7=0

Digit sum for option 4 = 2+3+5+6+7+8= 31= 3+1=4

For addition to be correct

Digit sum of all numbers in L.H.S = Digit sum of Answer in R.H.S.

Therefore, Answer is A.= 430388

We can solve it quickly with practice as below

Digit sum of 54326 + 32696 + 92765 + 85678 + 72567 + 92356

= (5+4+3+2+6) + (3+2+6+9+6) + (9+2+7+6+5) + (8+5+6+7+8) + (7+2+5+6+7) + (9+2+3+5+6)

= 2+8+2+7+0+7

= 26

= 2+6

= 8

practice will reduce time required for solving such questions (eliminate 9 , multiple of 9 or sum of digits as 9 with Zero).

[We will discuss latter that Any numbers which have digit sum=0, will be always divisible by 9].

?	**Check the addition whether correct or not by digit sum method.** **a. 535 + 275 = 810** **b. 195 + 105= 210** **c. 2565 + 3434 = 5989** **d. 956+ 132 = 1088** **e. 357+ 342 = 699** **f. 78677 + 11221 = 89878**

Addition of Number by converting Higher digit to lower digit with a bar raised for "subtraction".

To convert higher digit to lower as bar	*Example - Addition*
$-1=\bar{1}$ (Ex. $19=20-1=2\bar{1}$)	$1\bar{1}+1\bar{2}=2\bar{3}=20-3=17$
$-2=\bar{2}$ (Ex. $58=60-2=6\bar{2}$)	$23+2\bar{2}=23+20-2=41$
$-3=\bar{3}$ (Ex. $27=30-3=3\bar{3}$)	$29+28=3\bar{1}+3\bar{2}=6\bar{3}=57$

To convert higher digit to lower as bar	*Example - Addition*
$-4=\bar{4}$ (Ex. $26=30-4=3\bar{4}$)	$48+28=5\bar{2}+3\bar{2}=8\bar{4}=76$
$-5=\bar{5}$ (Ex. $25=30-5=3\bar{5}$)	$47+68=5\bar{3}+7\bar{2}=12\bar{5}=115$
$-6=\bar{6}$ (Ex. $24=30-6=3\bar{6}$)	$76+48=8\bar{4}+5\bar{2}=13\bar{6}=124$
$-7=\bar{7}$ (Ex. $23=30-7=3\bar{7}$)	$46+67=5\bar{4}+7\bar{3}=12\bar{7}=113$
$-8=\bar{8}$ (Ex. $22=30-8=3\bar{8}$)	We use this application for converting higher digit in addition or subtraction to lower digit with a bar as sign of subtraction. Ex. $19=2\bar{1}$, this way addition or subtraction becomes easy.
$-9=\bar{9}$ (Ex. $21=30-9=3\bar{9}$)	

?	**Add following by converting higher digit to lower.**		
	a. 29 + 31 =	**b. 48 + 59 =**	**c. 89 + 58 =**
	d. 19+ 52 =	**e. 615+389 =**	**f. 77 + 127 =**

CHAPTER 3
Subtraction

In school, we had been taught subtraction by borrowing method.

But here we will learn subtraction in five methods.

1. Borrowing method
2. Subtraction from left to right
3. Carry over method
4. Alteration of number
5. Digit sum method
6. Subtraction by Converting Higher Digit to Lower Digit

Borrowing Method

In our childhood, we all have studied this method as shown below.

For example

> Subtract 235 from 542,
>
> | | | | |
> |---|---|---|---|
> | | 5 | 4 | 2 |
> | − | 2 | 3 | 5 |
> | | 3 | 0 | 7 |
>
> We cannot subtract 5 from 2 so we borrow 10 from tens digit (4) and add 10 to unit number (2) and subtract 1 from tens number (4)
>
> So it becomes like
>
> | | | | |
> |---|---|---|---|
> | | 5 | 3 | (12) |
> | − | 2 | 3 | 5 |
> | | 3 | 0 | 7 |

Have we ever given a thought to why do we only borrow 10 and not any other number during this method?.

Let's understand this.

542 can be written as 500+ 40 +2 or it can also be written as 500 + 30 +12

542 = 500 + 40 + 2 = 5 4 2 [5 at hundred place, 4 at ten place , 2 at unit place]

542 = 500 + 30 + 12 = 5 3 (12) [5 at hundred place, 3 at ten place, 12 being considered at unit place]

Subtraction from Left to Right

In this method, we will subtract numbers from left to write.

Subtract 2517 from 5426,

$$\begin{array}{rcccc} & 5 & 4 & 2 & 6 \\ - & 2 & 6 & 1 & 7 \\ \hline \end{array}$$

We will start subtraction from left to right.

5 – 2 = 3, we will put 2 (1 less from 3) because digit at hundred's place is 4 in top row which is less than digit 6 at hundred's place in second row. Now put 1 near 4 to consider as 14.

$$\begin{array}{rcccc} & 5 & {}^{1}4 & 2 & 6 \\ - & 2 & 6 & 1 & 7 \\ \hline & 2 & & & \end{array}$$

Now 14 – 6 = 8, here we will put 8 below as 2 is greater than 1.

$$\begin{array}{rcccc} & 5 & {}^{1}4 & 2 & 6 \\ - & 2 & 6 & 1 & 7 \\ \hline & 2 & 8 & & \end{array}$$

Now 2 – 1 = 1, we will put 0 (1 less from 1) because 6 is less than 7. We will put 1 near 6 to consider it as 16.

$$\begin{array}{rcccc} & 5 & {}^{1}4 & 2 & {}^{1}6 \\ - & 2 & 6 & 1 & 7 \\ \hline & 2 & 8 & 0 & \end{array}$$

Now 16 – 7 = 9

$$\begin{array}{rcccc} & 5 & {}^{1}4 & 2 & {}^{1}6 \\ - & 2 & 6 & 1 & 7 \\ \hline & 2 & 8 & 0 & 9 \end{array}$$

?	**Subtract following by subtracting from left to right method.**
	a. 327 – 318 = **b. 6548 – 3339 =**
	c. 72567 – 52458 = **d. 96785 – 85668 =**
	e. 615 – 389 = **f. 677 – 187 =**

Carry Over Method

In this method, we carry over numbers similar to that of addition.

Subtract 235 from 542,

$$\begin{array}{cccc} & 5 & 4 & (10+2) \\ - & 2 & (3+1) & 5 \\ \hline & 3 & 0 & 7 \end{array}$$

We cannot subtract 5 from 2 so we add 10 to digit at units place(2) and add 1 to digit at tens place (3) in number being subtracted

So it becomes like

$$\begin{array}{cccc} & 5 & 4 & (12) \\ - & 2 & 4 & 5 \\ \hline & 3 & 0 & 7 \end{array}$$

?	**Subtract following by carry over method.**
	a. 327 - 318 = **b. 6548 - 3339=**
	c. 72567 - 52458 = **d. 96785 - 85668 =**
	e. 615 - 389 = **f. 677 - 187 =**

Alteration of Numbers

Have you ever met someone, who is facing problems in subtraction? Sometimes students find difficulties in borrow method.

What do we mean by altering the numbers?

Well, let's understand this. If there are two numbers a and b(a>b) and their subtraction is c, then we can write

$$a - b = c$$

here adding or subtracting any number to both a and b does not change value of subtraction.

Say n is a number being added to both a and b

$$(a + n) - (b + n) = a + n - b - n = a - b = c$$

Or if n being subtracted from both a and b then

$$(a - n) - (b-n) = a - n - b + n = a - b = c$$

Still the result does not change.

This way we can alter the numbers to nearest number of multiple of 10, 100..etc.

Let's see an example

Subtract 235 from 542,

$$\begin{array}{r} 5\ 4\ 2 \\ -\ 2\ 3\ 5 \\ \hline \end{array}$$

We cannot subtract 5 from 2 so we can add 5 to both numbers So that these numbers will become 240 and 547

$$\begin{array}{r} 5\ 4\ 7 \\ -\ 2\ 4\ 0 \\ \hline 3\ 0\ 7 \end{array}$$

Thus we see that result does not change.

We can see another example

Subtract 2356 from 5423,

$$\begin{array}{r} 5\ 4\ 2\ 3 \\ -\ 2\ 3\ 5\ 6 \\ \hline \end{array}$$

We cannot subtract 6 from 3 and 5 from 2 so we can add 4 to both numbers to alter the numbers as 2360 and 5427

$$\begin{array}{r} 5\ 4\ 2\ 7 \\ -\ 2\ 3\ 6\ 0 \\ \hline \end{array}$$

Now can again add 40 to both numbers to alter these as 2400 and 5467

$$\begin{array}{r} 5\ 4\ 6\ 7 \\ -\ 2\ 4\ 0\ 0 \\ \hline 3\ 0\ 6\ 7 \end{array}$$

This will reduce calculation time and make the subtraction easy without carry over or borrowing.

?

Subtract following by alteration method.

a. 327 - 318 =	**b. 6548 - 3339=**
c. 72567 - 52458 =	**d. 96785 - 85668 =**
e. 615 - 389 =	**f. 677 - 187 =**

Digit Sum Method for Subtraction

Subtraction is just reverse of addition and therefore if we consider a and b (a>b) and their subtraction as c then

$$a - b = c$$

or

a = b + c

therefore, Digit sum of a = Digit sum of (b + c)

if we consider above example to check whether **subtraction of** 2356 from 5423 is correct or not then

5423 – 2356 = 3067

Or

5423 = 2356 + 3067

Digit sum of 5423 = (5+4+2+3)= 0+2+3=5

Digit Sum of (2356 + 3067) = 2 + 3 + 5 + 6 + 3 + 0 + 6 + 7 =10+0+0+6+7=23=2+3=5

Therefore, Digit sum of 5423 = Digit Sum of (2356 + 3067)

students can check their answers and can also choose the correct multiple answer among the options by this method easily and quickly.

?	**Check by digit sum method whether following subtraction correct or not.** **a. 327 – 318 = 9** **b. 6548 – 3339 =3309** **c. 72567 – 52458 = 20109** **d. 96785 – 85668 = 11127** **e. 615 – 389 = 326** **f. 677 – 187 = 480**

Subtraction by Converting Higher Digit to Lower Digit with Bar as Subtraction

To convert higher digit to lower as bar	*Example-Subtraction*
$-1=\bar{1}$ (Ex. $19=20-1=2\bar{1}$)	$1\bar{2}-1\bar{1} = \bar{1} = -1$, $1\bar{1}>1\bar{2}$
$-2=\bar{2}$ (Ex. $58=60-2=6\bar{2}$)	$23-2\bar{2}=23-20+2=5$
$-3=\bar{3}$ (Ex. $27=30-3=2\bar{3}$)	$61-29=61-3\bar{1}=31-\bar{1}=32$
$-4=\bar{4}$ (Ex. $26=30-4=3\bar{4}$)	$5\bar{2}-3\bar{2}=20$
$-5=\bar{5}$ (Ex. $25=30-5=3\bar{5}$)	$68-47=7\bar{2} - 5\bar{3}=20-\bar{1}=21$
$-6=\bar{6}$ (Ex. $24=30-6=3\bar{6}$)	$76-48=8\bar{4}-5\bar{2}=3\bar{2}=28$
$-7=\bar{7}$ (Ex. $23=30-7=3\bar{7}$)	$67-46= 7\bar{3}-5\bar{4}=20-\bar{1}=21$
$-8=\bar{8}$ (Ex. $22=30-8=3\bar{8}$)	We use this application for converting higher digit in addition or subtraction to lower digit with a bar as sign of subtraction. Ex. $19=2\bar{1}$, this way addition or subtraction becomes easy.
$-9=\bar{9}$ (Ex. $21=30-9=3\bar{9}$)	

?	**Subtract by converting higher digit to lower digit.** **a. 37 – 18 =** **b. 120 – 59=** **c. 149 – 75=** **d. 650 – 147 =** **e. 54 – 39 =** **f. 76 – 29 =**

CHAPTER 4

Multiplication

So far we learn about different methods for Addition and Subtraction. Most students feel multiplication to be harder than addition or subtraction.

Is multiplication really that difficult?

Well, let's start with the method which we all have learnt in school but we will further simplify it in much easier steps so as to make it faster for us to calculate and reach the answer.

Carry Over Method of Multiplication

This method was taught to all of us during our childhood days. Almost all of us have learnt multiplication table from 1 to 9 as minimum requirements or few learnt more upto 20, 30 or even practice to remember as many as possible.

Table -1			
1	×	1	1
1	×	2	2
1	×	3	3
1	×	4	4
1	×	5	5
1	×	6	6
1	×	7	7
1	×	8	8
1	×	9	9
1	×	10	10

Table -2			
2	×	1	1
2	×	2	4
2	×	3	6
2	×	4	8
2	×	5	10
2	×	6	12
2	×	7	14
2	×	8	16
2	×	9	18
2	×	10	20

Table -3			
3	×	1	3
3	×	2	6
3	×	3	9
3	×	4	12
3	×	5	15
3	×	6	18
3	×	7	21
3	×	8	24
3	×	9	27
3	×	10	30

Table -4			
4	×	1	4
4	×	2	8
4	×	3	12
4	×	4	16
4	×	5	20
4	×	6	24
4	×	7	28
4	×	8	32
4	×	9	36
4	×	10	40

Table -5			
5	×	1	5
5	×	2	10
5	×	3	15
5	×	4	20
5	×	5	25
5	×	6	30
5	×	7	35
5	×	8	40
5	×	9	45
5	×	10	50

Table -6			
6	×	1	6
6	×	2	12
6	×	3	18
6	×	4	24
6	×	5	30
6	×	6	36
6	×	7	42
6	×	8	48
6	×	9	54
6	×	10	60

Table -7			
7	×	1	7
7	×	2	14
7	×	3	21
7	×	4	28
7	×	5	35
7	×	6	42
7	×	7	49
7	×	8	56
7	×	9	63
7	×	10	70

Table -8			
8	×	1	8
8	×	2	16
8	×	3	24
8	×	4	32
8	×	5	40
8	×	6	48
8	×	7	56
8	×	8	64
8	×	9	72
8	×	10	80

Table -9			
9	×	1	9
9	×	2	18
9	×	3	27
9	×	4	36
9	×	5	45
9	×	6	54
9	×	7	63
9	×	8	72
9	×	9	81
9	×	10	90

Table -10			
10	×	1	10
10	×	2	20
10	×	3	30
10	×	4	40
10	×	5	50
10	×	6	60
10	×	7	70
10	×	8	80
10	×	9	90
10	×	10	100

Generally we need to remember upto 9 or remembering more helps you calculate faster but remaining can be derived from 1 to 9 tables.

Table -11			
11	×	1	11
11	×	2	22
11	×	3	33
11	×	4	44
11	×	5	55
11	×	6	66
11	×	7	77
11	×	8	88
11	×	9	99
11	×	10	110

Table -12			
12	×	1	12
12	×	2	24
12	×	3	26
12	×	4	48
12	×	5	60
12	×	6	72
12	×	7	84
12	×	8	96
12	×	9	108
12	×	10	120

Table -13			
13	×	1	13
13	×	2	26
13	×	3	39
13	×	4	52
13	×	5	65
13	×	6	78
13	×	7	91
13	×	8	104
13	×	9	117
13	×	10	130

Table -14			
14	×	1	14
14	×	2	28
14	×	3	42
14	×	4	56
14	×	5	70
14	×	6	84
14	×	7	98
14	×	8	112
14	×	9	126
14	×	10	140

Table -15			
15	×	1	15
15	×	2	30
15	×	3	45
15	×	4	60
15	×	5	75
15	×	6	90
15	×	7	105
15	×	8	120
15	×	9	135
15	×	10	150

Table -16			
16	×	1	16
16	×	2	32
16	×	3	48
16	×	4	64
16	×	5	80
16	×	6	96
16	×	7	112
16	×	8	128
16	×	9	144
16	×	10	160

Table -17			
17	×	1	14
17	×	2	34
17	×	3	51
17	×	4	68
17	×	5	85
17	×	6	102
17	×	7	119
17	×	8	136
17	×	9	153
17	×	10	170

Table -18			
18	×	1	18
18	×	2	36
18	×	3	54
18	×	4	72
18	×	5	90
18	×	6	108
18	×	7	126
18	×	8	144
18	×	9	162
18	×	10	180

Table -19			
19	×	1	19
19	×	2	38
19	×	3	57
19	×	4	76
19	×	5	95
19	×	6	114
19	×	7	133
19	×	8	152
19	×	9	171
19	×	10	190

Table -20			
20	×	1	20
20	×	2	40
20	×	3	60
20	×	4	80
20	×	5	100
20	×	6	120
20	×	7	140
20	×	8	160
20	×	9	180
20	×	10	200

Tricks and Tips for Simple Method to Write Table above 9

Student need not require to remember all the tables. Generally, if table upto 9 is known then others can be fond out very easily.

Here we will see some examples to understand.

To write table of any two digit Numbers, We follow the following rules.

(a) Just write down table for digit at tens place.

(b) Now write down table of digit at unit place nearby.

(c) Now we will write result by putting unit digit in (b) at extreme right and tens or above digits in (a) at extreme left.

(d) now we will add tens digit of (b) and unit digit of (a) and write it in between extreme right and extreme left.

E.g.: Write Table for 12

Table for 1 (a)	*Table for 2 (b)*	*Extreme left(tens digit for (a)*	*Sum of unit digit of (a) and Tens digit of (b)*	*Extreme Right(unit digit of (b)*	*Table for 12*
01	02	0	1+0=1	2	12
02	04	0	2+0=2	4	24
03	06	0	3+0=3	6	36
04	08	0	4+0=4	8	48
05	10	0	5+1=6	0	60
06	12	0	6+1=7	2	72
07	14	0	7+1=8	4	84

Table for 1 (a)	*Table for 2 (b)*	*Extreme left(tens digit for (a)*	*Sum of unit digit of (a) and Tens digit of (b)*	*Extreme Right(unit digit of (b)*	*Table for 12*
08	16	0	**8+1=9**	6	96
09	18	0	**9+1=10**	8	108
10	20	1	**0+2=2**	0	120

Write table for 26

Table for 2 (a)	*Table for 6 (b)*	*Extreme left(tens digit for (a)*	*Sum of unit digit of (a) and Tens digit of (b)*	*Extreme Right(unit digit of (b)*	*Table of 26*
02	06	0	**2+0=2**	6	26
04	12	0	**4+1=5**	2	52
06	18	0	**6+1=7**	8	78
08	24	0	**8+2=10**	4	104
10	30	1	**0+3=3**	0	130
12	36	1	**2+3=5**	6	156
14	42	1	**4+4=8**	2	182
16	48	1	**6+4=10**	8	208
18	54	1	**8+5=13**	4	234
20	60	2	**0+6=6**	0	260

Here 208 written as (1extreme left digit of a+1 tens digit of 10) 08 and 234 as (1extreme left digit of a+1 tens digit of 13) 34

Now we have been taught carry over method of multiplications. For example:

If we need to multiply 8 × 9 then we simply write

8 × 9 7 2	(a) we multiply 9 × 8 from multiple tables and write 72 below as answer

Or if we need to multiply 78 x 9 then we multiply as

7 8 × 9 ——— 7 0 2	(a) We multiply 9 x 8 from multiple table and get 72 but we write 2 at unit place and carry 7 for tens place (b) Then multiply 9 x 7 and get 63, now add 63 + 7 =70 and write at tens place.

Or if we need to multiply 78 x 69 then we multiply as

7 8 × 6 9 ——— 7 0 2 4 6 8 ——— 5 3 8 2	(a) We multiply 9 x 8 from multiple table and get 72 but we write 2 at unit place and carry 7 for tens place in first row starting from unit place. (b) Then multiply 9 x 7 and get 63, now add 63 + 7 = 70 and write at tens place in first row. (c) Then multiply 6 x 8 and get 48 , but we write 8 at tens place in second row and carry 4 for hundred place in second row. (d) Then multiply 6 x 7 and get 42, now add 42 + 4 = 46 and write at hundred place in second row. (e) Now add numbers in each column starting from unit place and write 2 at unit place. 8 (8+0=8) at tens place, 3 (7 + 6 = 13, write 3 and carry 1), 5 (4 + 1 (carry over)) at thousand place.

This way we all have learnt multiplication in school by multiplying the numbers from right to left and then carrying over numbers and then adding, but it has sometimes proved to be confusing for us even in schooldays and sometimes afterwards too.

Here we will learn some more methods for quick multiplication.

Currency Count Method and Split and Merge Method for Multiplication

Previously we have learnt how to add from left to write as currency counting method for addition.

Now let's see examples. Multiply 578 x 69

Here 69 = 60 + 9

Therefore, by **split and merge method**

578 x 69 = 578 x (60 + 9)

= 578 x 60 + 578 x 9 = 34680 + 5202 = 39882

Multiply 542 x 23 = 542 x (20 + 3)

= 542x20 + 542 x3

= 10840 + 1626 = 12466

Multiply **578 x 469 = 578 x (400 + 60 + 9)**

= 578 x 400 + 578 x 60 + 578 x 9

= 231200 + 34680 + 5202

= 271082

Isn't it simple to multiply for any order of multiplication through this method?

Let's multiply 3578 x 2469

3578 x 2469 = 3578 x (2000 + 400 + 60 + 9)

= 7156000 + 1431200 + 214680 + 32202

= 8834082

We split numbers as per our convenience of multiplication

Multiply 78 x 69 = 78 x (60 + 9) = 4680 + 702 = 5382

Multiply 49 x 51 = 49 x (50 + 1) = 2450 + 49 = 2499

Multiply 41 x 26 = (40+1) x 26 = 1040 + 26 = 1066

Multiply 35 x 61 = 35 x (60 + 1) = 2100 +35 = 2135

Multiply 71 x 43 = (70 + 1) x 43 = 3010 + 43 = 3053

Multiply 86 x 11 = 86 x (10 + 1) = 860 + 86 = 946

?	**Multiply by split and merge method.**		
	a. 35 x 21 =	**b. 120 x 42 =**	**c. 132 x 51 =**
	d. 650 x 12 =	**e. 54 x 53 =**	**f. 76 x 31 =**
	g. 52 x 14 =	**h. 23 x 12 =**	**i. 49 x 21 =**

Multiplication based on $(a + b)(a - b) = a^2 - b^2$

In any multiplication of two numbers having numbers close to each other by 2,4,6, 8....etc. is calculated by using formula.

$$(a + b)(a - b) = a^2 - b^2$$

Where a and b are any two numbers

Let's see some example

Multiply 29 x 31, here 31-29=2

29 x 31 = (30-1)x(30+1) = 302 – 12 = 900 – 1 = 899

Multiply 28 x 32, here 32-28=4

28 x 32 = (30-2)x(30+2) = 302 – 22 = 900 – 4 = 896

Multiply 34 x 36, here 36-34=2

34 x 36 = (35-1)x(35+1) = 352 – 12 = 1225 – 1 = 1224

Multiply 48 x 52, here 52-48=4

48 x 52 = (50-2)x(50+2) = 502 – 22 = 2500 – 4 = 2496

Multiply 99 x 101, here 101-99= 2

99 x 101 = (100-1)x(100+1) = 1002 – 12 = 10000 – 1 = 9999

?	**Multiply by using method (a+b)(a-b)= $a^2 - b^2$**		
	a. 33 x 27 =	**b. 19 x 21=**	**c. 18 x 22 =**
	d. 39 x 41 =	**e. 49 x 51 =**	**f. 56 x 64 =**
	g. 53 x 47 =	**h. 23 x 17 =**	**i. 69 x 71 =**

Multiplication of numbers having same base digit at tens place and sum of digits at units place is 10.

If XZ and XY are two numbers such that Z+Y=10 then

Multiplication of XZ and XY is represented as by numbers obtained from multiplication of X and X+1 and multiplication of Z and Y.

Let's see some example

Multiply 12 x 18, here 1 is at tens place in both number and 2+8=10

12x18 = 216, (Here 1x(1+1)= 1 x 2 =2 and 2 x 8 =16)

Multiply 13 x 17, here 1 is at tens place in both number and 3+7=10

13x17 = 221, (Here 1x(1+1)= 1 x 2 =2 and 3 x 7 =21)

Multiply 14 x 16, here 1 is at tens place in both number and 4+6=10

14x16 = 224, (Here 1x(1+1)= 1 x 2 =2 and 4 x 6 =24)

Multiply 15 x 15, here 1 is at tens place in both number and 5+5=10

15x15 = 152 =225, (Here 1x(1+1)= 1 x 2 =2 and 5 x 5 =25)

Multiply 22 x 28, here 2 is at tens place in both number and 2+8=10

22x28 = 616, (Here 2x(2+1)= 2 x 3 =6 and 2 x 8 =16)

Multiply 33 x 37, here 3 is at tens place in both number and 3+7=10

33x17 = 1221,(Here 3x(3+1)= 3 x 4 =12 and 3 x 7 =21)

Multiply 44 x 46, here 4 is at tens place in both number and 4+6=10

44x46 = 2024, (Here 4x(4+1)= 4 x 5 =20 and 4 x 6 =24)

Multiply 45 x 45 , here 4 is at tens place in both number and 5+5=10

45x45 = 452 =2025 , (Here 4x(4+1)= 4 x 5 =20 and 5 x 5 =25)

?

Multiply following:

a. 31 x 39 =	**b. 19 x 11 =**	**c. 18 x 12 =**
d. 49 x 41 =	**e. 59 x 51 =**	**f. 56 x 54 =**
g. 53 x 57 =	**h. 23 x 27 =**	**i. 69 x 61 =**

Multiplication of numbers having same base digit at tens place and sum of digits at units place may be less than 10.

Let's see some example

Multiply numbers having tens place as 1,

14 x 1**3** = 10 x (**14** + **3**) + 4 x 3 = 10 x 17 + 12 = 170 +12 = 182

14 x 1**5** = 10 x (**14** + **5**) + 4 x 5 = 10 x 19 + 20 = 190 +20 = 210

16 x 17= 10 x 23 + 42 = 230 + 42 = 272

18 x 17 = 10 x 25 + 56 = 250 + 56 = 306

Multiply numbers having tens place as 2

24 x 2**7** = 20 x (**24** + **7**) + 4 x 7 = 20 x 31 + 28 = 620 + 28 = 648

23 x 2**8** = 20 x (**23** + **8**) + 3 x 8 = 20 x 31 + 24 = 620 + 24 = 624

21 x 25 = 20 x 26 + 5 = 520 + 5 = 525

22 x 26 = 20 x 28 + 12 = 560 + 12 = 572

24 x 28= 20 x 32 + 32 = 640 + 32 = 672

Multiply numbers having tens place as 3

34 x 37 = 30 x (**34** +7) + 4 x 7 = 30 x 41 + 28 = 1230+28=1258

35 x 39 = 30 x (35 + 9) + 5 x 9 = 30 x 44 + 45 = 1320+45=1365

Similarly

43 x 49 = 40 x 52 + 27 = 2080+27=2107

56x 57= 50 x 63 + 42 – 3150 + 42 = 3192

68 x 69 = 60 x 77 + 72 = 4620 + 72 = 4692

75 x 79 = 70 x 84 + 45 = 5880 + 45 = 5925

84 x 83 = 80 x 87 + 12 = 6960 +12 = 6972

98 x 99 = 90 x 107 + 72 =9630 + 72 = 9702

?	**Multiply following:**		
	a. 31 x 38 =	**b. 19 x 16 =**	**c. 18 x 14 =**
	d. 49 x 43 =	**e. 59 x 56 =**	**f. 56 x 52 =**
	g. 53 x 54 =	**h. 23 x 28 =**	**i. 69 x 66 =**
	j. 86 x 88 =	**k. 45 x 47 =**	**l. 92 x 94 =**

Squaring of Numbers

Squaring a number ending with 5.

Square of any number ending with 5 can be represented by writing 25 and number obtained by multiplication of digit at tens place with one digit higher than the tens place.

Let's see some example

15^2 = 225 (1 x (1+1) = 1 x 2 = 2 and 25)

25^2 = 625 (2 x (2+1) = 2 x 3 = 6 and 25)

35^2 = 1225 (3 x (3+1) = 3 x 4 = 6 and 25)

45^2 = 2025 (4 x (4+1) = 4 x 5 = 20 and 25)

55^2 = 3025 (5 x (6+1) = 5 x 6 = 30 and 25)

65^2 = 4225 (6 x (6+1) = 6 x 7 = 42 and 25) and so on

Squaring a Number

Square of any number can be obtained as follow:

Let's see some example

14^2 = 10 x (14 + **4**) +**4x4** = 10 x 18 +16 = 196

24^2 = 20 x (24 + **4**) + **4x4** = 560 + 16 = 576

34^2 = 30 x (34 + **4**) + **4x4** = 1140 + 16 = 1156

$18^2 = 10 \times (18 + 8) + 64 = 260 + 64 = 324$
$17^2 = 10 \times (17 + 7) + 49 = 240 + 49 = 289$
$25^2 = 20 \times (25 + 5) + 25 = 600 + 25 = 625$
$28^2 = 20 \times (28 + 8) + 64 = 720 + 64 = 784$
$37^2 = 30 \times (37 + 7) + 49 = 1320 + 49 = 1369$
$44^2 = 40 \times (44 + 4) + 16 = 1920 + 16 = 1936$
$56^2 = 50 \times (56 + 6) + 36 = 3100 + 36 = 3136$
$84^2 = 80 \times 88 + 16 = 7040 + 16 = 7056$
This method is as same as we found multiplication of numbers having same digit at tens place
$17^2 = 17 \times 17 = 10 \times (17 + 7) + 49 = 240 + 49 = 289$
$28^2 = 28 \times 28 = 20 \times (28 + 8) + 64 = 720 + 64 = 784$
$37^2 = 37 \times 37 = 30 \times (37 + 7) + 49 = 1320 + 49 = 1369$

?	**Find square of following numbers:**		
	a. $31^2 =$	**b.** $19^2 =$	**c.** $18^2 =$
	d. $49^2 =$	**e.** $59^2 =$	**f.** $56^2 =$
	g. $53^2 =$	**h.** $23^2 =$	**i.** $69^2 =$
	j. $86^2 =$	**k.** $45^2 =$	**l.** $92^2 =$

You would think its over now....but not yet...let's see some more examples and more methods.

Multiply by Alteration of Numbers

Doubling of numbers

67 x 2 = 134 [just double the number]

67 x 4 = 67 x 2 x 2 = 134 x 2 = 268 [double twice]

67 x 8 = 67 x 2 x 2 x 2 = 134 x 2 x 2 = 268 x 2 = 536 [double thrice]

67 x 16 = 134 x 8 = 268 x 4 = 536 x 2 = 1072 [double four times]

67 x 32 = 134 x 16 = 268 x 8 = 536 x 4 = 1072 x 2 = 2144 [double five times]

Doubling and tripling

67 x 6 = 67 x 2 x 3 = 134 x 3 = 402

67 x 12 = 134 x 2 x 3 = 268 x 3 = 804

67 x 24 = 134 x 12 = 268 x 6 = 536 x 3 = 1608

67 x 48 = 134 x 24 = 268 x 12 = 536 x 6 = 1072 x 3 = 3216

Multiplying by 5 , 25 , 50 etc.

Here 5 can be written as 10/2 , 25 can be written as 100/4, 50 can be written as 100/2 and so on.

Therefore

67 x 5 = (67 x 10) /2 = 670 / 2 = 335

67 x 50 = [67 x 100]/2 = 6700/2= 3350

67 x 25 = [67 x 100]/4=6700/4=3350/2=1675

Or 67 x 25 = [67x50]/2=3350/2=1675

?	**Solve following multiplications:**		
	a. 34 x 4 =	**b. 17 x 6 =**	**c. 18 x 8 =**
	d. 49 x 5 =	**e. 58 x 50 =**	**f. 56 x 25 =**
	g. 27 x 25 =	**h. 28 x 6 =**	**i. 69 x 50 =**
	j. 86 x 5 =	**k. 64 x 25 =**	**l. 84 x 50 =**

Multiply any number by 11

Multiplying any number with 11 is very tricky yet quick.

To multiply any number by 11, write down first digit of number then add digits in between and carry over for tens, etc. and lastly write the last digit of the number.

Now let's see some example.

12 x 11 = 1 / **1+2** / 2 = 132
82 x 11 = 8/ **8+2**/2 = 8/10/2=902
(Carry one for next digit and put 0 at tens place, here 8+1 = 9 at hundred place)

98 x 11 = 9/**9+8**/8= 9/17/8=1078
(Carry one for next digit and put 7 at tens place, here 9+1=10 at hundred place)

987 x 11 = 9/ **9+8/8+7**/7= 9/17/15/7=10857
9876 x 11 = 9/9+8/8+7/7+6/6= 9/17/15/13/6= 108636

5298764 x 11 = 5 /5 + 2/ 2 + 9/9 + 8/8 + 7/7 + 6/6 + 4/4 = 5/7/11/17/15/13/10/4= 58286404

You may directly write as follow
325678 x 11 = 3/5/7/11/13/15/8 = 3582458

?	**Solve following multiplications:**		
	a. 34 x 11 =	**b. 176 x 11 =**	**c. 2618 x 11 =**
	d. 567 x 11 =	**e. 558 x 11 =**	**f. 256 x 11 =**
	g. 273 x 11 =	**h. 28 x 11 =**	**i. 69 x 11 =**
	j. 45686 x 11 =	**k. 678964 x 11 =**	**l. 823584 x 11 =**

Now let's see , how to multiply any number with 111 or 1111 or 11111 etc...

Since 111= 100 + 11,

1111= 1000 + 111 = 1000 + 100 + 11

Therefore,

Multiplying any number with 111

= multiply any number with (100 + 11)

= multiply the number with 100 + multiply the number with 11.

Let's see some example

12 x 111 = 12 x (100 + 11) = 1200 + 12 x 11 = 1200 + 132= 1332 or

12 x 111 = 1 / **1+2/1+2**/2=1/3/3/2= 1332

[for any two digit number to multiply with 111]

12 x 1111 = 12 (1000 + 100 +11) = 12000 + 1200 + 132 = 13332 or

12 x 1111= 1/**3/3/3**/2 = 13332

[for any two digit number to multiply with 1111]

12 x 11111= 1/**3/3/3/3**/2= 133332

[for any two digit number to multiply with 11111]

82 x 111 = 8200 + 82 x 11 = 8200 + 902 = 9102 Or

82 x 111 = 8 / **8+2 / 8+2** / 2 = 8/10/10/2=9102

[for any two digit number to multiply with 111]

82 x 1111 = 82 x (1000 +100+11)= 82000 + 8200 + 902
= 91102 or

82 x 1111= 8/**10/10/10**/2= 91102

[for any two digit number to multiply with 1111]

98 x 111 = 9/**17/17**/8= 10878

98 x 1111 = 9/**17/17/17**/8= 108878

198 x 111 = 19/**19+8/19+8**/8= 19/**27/27**/8= 21978

[for any three digit number to multiply with 111]

198 x 1111= 19/**27/27/27**/8 = 219978

576 x 111 = 57/**63/63**/6= 63936

576 x 1111= 57/**63/63/63**/6 = 639936

?	**Solve following multiplications:**		
	a. 34 x 111 =	**b. 176 x 111 =**	**c. 28 x 1111 =**
	d. 567 x 11 =	**e. 58 x 1111 =**	**f. 256 x 11 =**
	g. 27 x 111 =	**h. 28 x 11 =**	**i. 69 x 11 =**
	j. 86 x 111 =	**k. 64 x 1111 =**	**l. 82 x 11 =**

Thus any number multiplied by 11 can be obtained easily with the above methods and we can also use similar methods for multiplying two or three digit numbers with 111, 1111 etc.. while four or five digits numbers or above can be multiplied by dissociating 111 , 1111 into their bases (1000, 100 , 10 etc.)

So let's see whether we can use this method to multiply any number with 22 , 33 , 44, 55, 66, 77 , 88, 99, 333, 444, 555 etc.

Let's See some example

12 x 22 = 12 x 2 x 11 = 24 x 11 = 264

82 x 22 = 82 x 2 x 11 = 164 x 11 = 1/7/10/4= 1804

98 x 33 = 294 x 11 = 2/11/13/4= 3234

768 x 44 = 3072 x 11 = 3/3/7/9/2=33792

Now let's see few more examples like

12 x 222 = 24 x 111= 2/6/6/4=2664

82 x 222 = 164 x 111= 16/20/20/4= 18204

98 x 333 = 294 x 111 = 29/33/33/4 = 32634

98 x 3333= 294 x 1111= 29/33/33/33/4= 326634

So this method to multiply numbers by changing them in multiples of 11, etc. will surely help to multiply faster.

?	**Solve following multiplications:**		
	a. 34 x 22 =	**b. 25 x 333 =**	**c. 23x 444 =**
	d. 132 x 33 =	**e. 58 x 55 =**	**f. 234 x 222 =**
	g. 27 x 333 =	**h. 28 x 66 =**	**i. 69 x 99 =**
	j. 86 x 77 =	**k. 35 x 44 =**	**l. 82 x 33 =**

Multiply Two Numbers Near Base

(Base of any numbers are their nearest 10, 100, 1000, 10000 ...etc.)

Now just check how much time you take to multiply 996 x 998.

Now we will do it quickly.

We consider base number as 100 for numbers close to it like 96, 94 , 98 ....102, 103 , 105 .....etc.

Similarly, we consider base number as 1000 for numbers close to it like 998, 997, 996....1001,1002, 1004... etc. and so on for 10000, 100000...etc.

Now we check out how much smaller or larger the numbers are from their nearest base numbers.

And write that value with –ve sign if smaller and +ve sign if larger.

For example: Multiply 98 x 96

So, 98 is lower by 2 from it's nearest base number 100 and 96 is lower by 4 from it's nearest base number.

Step 1: Now write 98 and -2 with some space between them

98	–2

Step 2: Now write 96 and -4 below it in same way,

98	–2
96	–4

Step 3: Multiply (2) and (-4) which is 8 and put it down below as 08 due to base 100(**two digit for base 100**)

98	–2
96	–4
	08

Step 4: Add diagonally, you will gate 94, put 94 and the final answer is 9408.

98	−2
96	−4
94	08

Let's see some more example.

Multiply 97 x 95

97	−3	Difference from base number 100 is −3
95	−4	Difference from base number 100 is −5
92	15	Product of (−3) and (−5) is 15
		Diagonal addition is 92

97 x 95 = 9215

Multiply 103 x 95

		Difference from base number 100 is 3
103	3	Difference from base number 100 is −5
95	−5	Product of (3) and (−5) is −15
97	85	Diagonal addition is 98
		Now reduce 1 from 98 to add 100 to −15 so final

103 x 95 = 9785

Multiply 103 x 102

		Difference from base number 100 is 3
1 0 3	3	Difference from base number 100 is 2
1 0 2	2	Product of (3) and (2) is 6 so we will
1 0 5 0	6	write two digit number 06, Diagonal addition is 105, Now final answer is

103 x 102 = 10506

Multiply 998 x 996

9	9	8			-2	
9	9	7			-3	
9	9	5	0	0	6	

Difference from base number 1000 is –2
Difference from base number 100 is –3
Product of (–2) and (–3) is 6 write 3 digit number 006 for base 1000, Diagonal addition is 995, Now final answer is

998 x 996 = 995006

Multiply 998 x 1003

9	9	8				-2
1	0	0	3			3
1	0	0	0	9	9	4

Difference from base number 1000 is –2
Difference from base number 100 is 3
Product of (–2) and (3) is –6, write 3 digit number 006 for base 1000, Diagonal addition is 1001, Now reduce 1 from 1001 to add 1000 to (–6) so you get 1000 and 994, So final answer is

998 x 1003 = 1000994

Multiply 998 x 1005

9	9	8				-2
1	0	0	5			5
1	0	0	0	9	9	4

Difference from base number 1000 is –2
Difference from base number 100 is 5
Product of (–2) and (5) is –10, write 3 digit number 010 for base 1000, Diagonal addition is 1003, Now reduce 1 from 1003 to add 1000 to (–10) so you get 1002 and 990, So final answer is

998 x 1005 = 1002990

How to multiply 96 x 49 quickly?

96 x 49 is equal to 96 x (98)/2.

So we have 96 x 98 = 9408, therefore 96 x 49 = 96 x 98/2 = 4704

How to multiply 996 x 497 quickly?.

996 x 497 is equal to 996 x (994)/2.

So we have 996 x 994 = 990024, therefore 996 x 497

= 996 x 998/2 = 495012

?	**Solve following multiplications:**		
	a. 97 x 96 =	**b. 99 x 96 =**	**c. 98x 99 =**
	d. 102 x 103 =	**e. 101 x 102 =**	**f. 99 x 103 =**
	g. 98 x 49 =	**h. 1002 x 1003 =**	**i. 1003 x 997 =**
	j. 999 x 998 =	**k. 1002 x 499 =**	**l. 1004 x 996 =**

Multiplication Check by Digit Sum Method

In any Multiplication there are three Terms: Multiplicand, Multiplier and Product.

In multiplication of two numbers,

digit sum of multiplicand x digit sum of multiplier = digit sum of product

For example: Multiply 99 x 98

We have seen earlier 99 x 98 = 9702

Digit Sum of Multiplicand = Digit Sum of 99 = 0

Digit Sum of Multiplier = Digit Sum of 98 = 8

Digit Sum of Product = Digit Sum of 9702 = 0

Product of Digit Sum of Multiplicand and Digit Sum of Multiplier = 0 x 8 = 0

Here, Product of Digit Sum of Multiplicand and Digit Sum of Multiplier=0 which is equal to digit sum of Product.

<table>
<tr><td colspan="2">Digit Sum of (Digit Sum of Multiplicand x Digit Sum of Multiplier) = Digit Sum of Product.
This is condition for any product of any two numbers to be correct.</td></tr>
<tr><td>?</td><td>Check following multiplications are correct or not by digit sum method:
a. 97 x 96 = 9312
b. 99 x 96 = 2524
c. 98 x 99 = 9712
d. 102 x 103 = 10506
e. 101 x 102 = 10202
f. 99 x 103 = 10197
g. 98 x 49 = 4502
h. 1002 x 1003 = 1005006
i. 1003 x 997 = 999791
j. 999 x 998 = 997002
k. 1002 x 499 = 496998
l. 1004 x 996 = 999984</td></tr>
</table>

CHAPTER 5
Division

Division is reverse of Multiplication however students feel it tough to understand.

Before learning about Division, We should discuss about numbers – Even, Odd, Composite, Prime.

Even Numbers: All the numbers which are divisible by 2 are called Even numbers.

E.g: 2,4,6,8,10,12,14,16,18.......etc.

Odd Numbers: Those numbers which are not divisible by 2 are called odd numbers.

Eg.: 1,3,5,7,9,11,13.........etc.

Prime Numbers: Those numbers which are only divisible by 1 and themselves

Eg:- 1,2,3,5,7,11,13,17,19.....etc.

Here 1 is only divisible by 1.

2 is only divisible by 1 or 2

3 is only divisible by 1 or 3, etc.

Composite Numbers: All the numbers which can be written in the form of factors are called composite numbers.

Eg.:- 4,6,8,9,10,12,14,15,16.....etc.

4 = 2 x 2

6 = 2 x 3

8 = 2 x 2 x 2 = 4 x 2

9 = 3 x 3

10 = 2 x 5

12 = 2 x 2 x 3 = 3 x 4 = 2 x 6 etc.

Divisibility Rule

Now we will understand divisibility rule:

1. How to identify a number divisible by 2?

   Any number ending with even digits such as 0, 2, 4, 6, 8, etc. are divisible by 2. This also means that all even numbers are divisible by 2.

   E.g.: 25678654 will be divisible by 2 because last digit is 4 which is divisible by 2.

   7689563 will not be divisible by 2 because last digit is 3 which is not divisible by 2.

876528 will be divisible by 2 because last digit is 8 which is divisible by 2.

try whether following numbers are divisible by 2 or not

(a) 67890

(b) 543289

(d) 9753218

(e) 3217860

(f) 7647275

2. How to identify a number divisible by 3?.

Any number will be divisible by 3 when digit sum of number is 0, 3 or 6.

E.g.:

(a) 2356789 has digit sum= 2+3+5+6+7+8+9

= 2+7+3+6+9+5+8=0+0+0+13=13=1+3=4

(Eliminate 9 or sum equal to 9 as 0)

Sine the digit sum is 4 therefore the number 2356789 not divisible by 3.

Now we can see that any combination of digits of this number 2356789 will not be divisible by 3 because the digit sum for all he combination will be 4.

Numbers with different combinations are 3256789, 5326789, 9523678, 8975236 ….etc have digit sum =4 and these numbers are not divisible by 3.

(b) 235746 has digit sum = 2+3+5+7+4+6 = 0+0+0=0, here digit sum is 0 and therefore number 235746 is divisible by 3.

Here we can see that any number formed from combination of digits of number 235746 will be divisible by 3 because the digit sum will be 0.

Numbers with different combinations are 352647, 745632, 753426....etc have digit sum =0 and therefore these numbers will be divisible by 3.

(c) 896754 has digit sum = 8+9+6+7+5+4= 8+0+6+7+0=21=2+1=3, here digit sum is 3 and therefore number 896754 is divisible by 3.

Here we see that any number formed from combination of digits of number 896754 will be divisible by 3 because the digit sum will be 3.

Numbers with different combinations are 457698, 675489, 968547....etc have digit sum =3 and therefore these numbers will be divisible by 3.

(d) 896757 has digit sum= 8+9+6+7+5+7= 8+0+6+7+5+7=33=3+3=6 here digit sum is 6 and therefore number 896757 is divisible by 3.

Here we can see that any number formed from combination of digits of number 896757 will be divisible by 3 because the digit sum will be 6.

Numbers with different combinations are 757698, 675789, 968577....etc have digit sum =6 and therefore these numbers will be divisible by 3

?	**Identify Number divisible by 3**		
	a. 9723 =	**b. 4296 =**	**c. 36577245 =**
	d. 102654 =	**e. 89676549 =**	**f. 6789345 =**
	g. 6789456 =	**h. 1002 =**	**i. 1563 =**
	j. 468 =	**k. 14949 =**	**l. 2346381 =**

3. How to identify a number divisible by 4?

(a) Any number having last two digit numbers divisible by 4 will be always divisible by 4.

E.g.: 276896 has last two digit number 96. Here 96 is divisible by 4 therefore the number 276896 is divisible by 4.

(b) Any number having last two digit numbers divisible by 2 have even remainder will be always divisible by 4.

E.g.: 276896 has last two digit number 96. Here 96 when divided by 2 has 48 as even remainder so 276896 is divisible by 4.

(c) Any number having last two digit (Say Q,R) is such that (QX2+R) is divisible by 4 then the number will be always divisible by 4.

E.g.: 276896 has last two digit 9 and 6. Here (QX2+R)=9X2+6=24 which is divisible by 4 so 276896 is divisible by 4.

(d) Any number having last two digit (Say Q,R) is such that (Q+R/2) is divisible by 2 then the number will be always divisible by 4.

(e) E.g.: 276896 has last two digit 9 and 6. Here (Q+R/2)=9+3=12 which is divisible by 2 so 276896 is divisible by 4.

?	**Identify Number divisible by 4**		
	a. 9724 =	**b. 4296 =**	**c. 36577248 =**
	d. 102616 =	**e. 89676544 =**	**f. 6789384 =**
	g. 6789436 =	**h. 1672 =**	**i. 1568 =**
	j. 468 =	**k. 14946 =**	**l. 2346382 =**

4. How to identify a number divisible by 5?.

Any Number ending with 0 or 5 will be divisible by 5.

E.g.: 205, 2345980, 4568905, ...etc., All these number ending with either 0 or 5 and therefore these numbers are divisible by 5.

?	**Identify Number divisible by 5**		
	a. 9725 =	**b. 4290 =**	**c. 36577250 =**
	d. 102600 =	**e. 8967615 =**	**f. 6789305 =**
	g. 6789455 =	**h. 1672 =**	**i. 1559 =**
	j. 4685 =	**k. 14946 =**	**l. 2346382 =**

5. How to identify a number divisible by 6?.

   Any number having digit sum is 0,3 or 6 and the number is even then the number will be divisible by 6.

   E.g.: Consider a number 5346798 having digit sum

   = 5+3+4+6+7+9+8

   =5+4+3+6+9+7+8

   =0+0+0+7+8=15=1+5=6

   Here the digit sum of the number is 6 and the number 5346798 is even therefore this number is divisible by 6.

   Here we can observe that any even number formed by rearranging digits of number 5346798 will be divisible by 6 because the even number formed will have digit sum = 6.

   Numbers 5798346, 5798364, 9753864, 5793648...... etc. are divisible by 6 because these numbers are even and having digit sum=6.

Number 534648 has digit sum= 5 + 3 + 4 + 6 + 4 + 8 = 0 + 0 + 4 + 8 = 12 = 1 + 2 = 3

Here the digit sum of the number is 3 and the number 5346748 is even therefore this number is divisible by 6.

Here we can observe that any even number formed by rearranging digits of number 5346748 will be divisible by 6 because the even number formed will have digit sum = 6.

Numbers 5748346, 5748364, 4753864, 5743648...... etc. are divisible by 6 because these numbers are even and having digit sum = 3.

?	**Identify Number divisible by 6**		
	a. 9723 =	**b. 4296 =**	**c. 367245 =**
	d. 102654 =	**e. 896769 =**	**f. 67932 =**
	g. 6789453 =	**h. 1002 =**	**i. 1566 =**
	j. 468 =	**k. 14949 =**	**l. 23412 =**

6. How to identify a number divisible by 7?

   There is no perfect way to identify divisibility of number with 7. Better and accurate way is to divide the complete number by 7 and check the remainder is zero or not.

7. How to identify a number divisible by 8?.

   (a) Any number will be divisible by 8 when last three digit number is divisible by 8.

E.g.: Number 8697457640 is divisible by 8 because last three digit number 640 is divisible by 8

(b) Any number will be divisible by 8 when last three digit number is divisible by 4 with even remainder.

E.g.: 8697457640 has last three digit number 640 which is when divided by 4 gives even remainder 160. Hence the number is divisible by 8.

(c) Any number will be divisible by 8 when last three digit number is divisible by 2, two times with even remainder.

E.g.: 8697457640 has last three digit number 640 which is when divided by 2 gives even remainder 320. When further divided by 2 gives even remainder 160. Hence the number is divisible by 8.

(d) Any number will be divisible by 8 when last three digit (Say P, Q, R) is such that (PX2+Q+R/2) is divisible by 2 giving even remainder Or divisible by 4.

E.g.: 8697457648 has last three digit number 648

Here (PX2+Q+R/2) = 6x2+4+4=20 which is divisible by 4 therefore the number 8697457648 is divisible by 8.

(e) Any number will be divisible by 8 when last three digit (Say P,Q,R) is such that (PX4+QX2+R) is divisible by 8.

E.g.: 8697457648 has last three digit number 648

Here (PX4+QX2+R)= 6x4+4X2+8=40 which is divisible by 8 therefore the number 8697457648 is divisible by 8.

?	**Identify Number divisible by 8**		
	a. 9712 =	**b. 4296 =**	**c. 36577240 =**
	d. 102618 =	**e. 8967600 =**	**f. 6789200 =**
	g. 6789434 =	**h. 1672 =**	**i. 1568 =**
	j. 468 =	**k. 14946 =**	**l. 2346382 =**

8. How to identify a number divisible by 9?

Any number will be divisible by 9 when digit sum of the number is 0.

E.g.: 6766839 has digit sum = 6+7+6+6+8+3+9= 6+ 7+6+8+6+3+9=6+7+6+8+0+0=27=2+7=9=0

(digit sum of 9=0)

Since the digit sum =0 therefore the number is divisible by 9.

?	**Identify Number divisible by 9**		
	a. 9729 =	**b. 45396 =**	**c. 36577244 =**
	d. 102654 =	**e. 89676549 =**	**f. 6189345 =**
	g. 6789456 =	**h. 6003 =**	**i. 5463 =**
	j. 45261 =	**k. 14949 =**	**l. 2346381 =**

9. How to identify a number divisible by 10?

   Any Number ending with 0 will be divisible by 10.

   E.g.: Number 7265390 ending with 0 therefore the number is divisible by 10.

10. How to identify a number divisible by 11?

    When sum of digits at even position of any number subtracted from sum of digits at odd position and the result is either 0, 11 or multiple of 11 then the number is divisible by 11.

    E.g.: 626835 has 6,6,3 on odd position and 2,8,5 on even position. Now sum of digits at odd place (6+6+3) – sum of digits at even place (2+8+5)

    =15-15=0, hence number is divisible by 11.

    In a number 9862320413

    (9+6+3+0+1)-(8+2+2+4+3)=19-19=0

    In a number 21901

    (2+9+1)-(1+0) = 12-1=11, hence the number is divisible by 11.

?	Identify Number divisible by 11		
	a. 275 =	b. 6479 =	c. 39446 =
	d. 76483 =	e. 5225 =	f. 6798 =
	g. 678942 =	h. 1298 =	i. 1463 =
	j. 4785 =	k. 14949 =	l. 2346311 =

11. How to identify a number divisible by 12?.

    12 has factors like 2x6, 2x2x3, 4x3, here we can see that any number will by divisible by 12 only when

    (a) The number is divisible by 12, if the number is divisible by 4 as well as divisible by 3.

        E.g.: 1068 has digit sum = 1 + 0 + 6 + 8 = 6 (eliminating 9 with 0), So the number is divisible by 3 and the last two digit number 68 is divisible by 4 hence the number 1068 is divisible by 4. Since 1068 is divisible by 3 as well as 4 therefore the number is divisible by 12.

?	Identify Number divisible by 12		
	a. 9726 =	b. 4296 =	c. 5496 =
	d. 54828 =	e. 67584 =	f. 3096 =
	g. 5472 =	h. 8376 =	i. 7032 =
	j. 46864 =	k. 6946 =	l. 5472 =

How to Identify Remainder

1. Remainder when any number divided by 2 will be always 0 or 1.

   E.g.: Any number 234567 has last digit odd and therefore remainder will be always 1.

   Any number 234568 has last digit even and therefore remainder will be always 0.

2. Remainder when any number divided by 3 will be either 1,2 or 0.

   Any number 214563 is divisible by 3 as the digit sum equal to three (2+1+4+5+6+3=3+9+9=3+0+0=3).

   Hence 3)214563 = 3)21/45/63 = 7/15/21=71521

   Any number 156325 has digit sum equal to 4 (1+5+6+3+2+5=6+9+7=13=4), hence remainder is 1.

   3)156325 = 3)156324 r 1. = 3)15/63/24 r 1= 5/21/08 r1 =52108 r1.

?

Identify Remainder when following number divided by 3

a. 9721 =	**b. 4296 =**	**c. 36577246 =**
d. 102656 =	**e. 89676549 =**	**f. 6789343 =**
g. 6789455 =	**h. 1004 =**	**i. 15634 =**
j. 4689 =	**k. 14949 =**	**l. 2346385 =**

3. Remainder when any number divided by 4 will be either 1,2,3 or 0.

   We shall follow divisibility rule of any number by 4.

   E.g.: Any number 124865 will have remainder 1, when divided by 4 because last two digit number 65 is odd which is not divisible by 4. But 64 smaller than 65 is divisible by 4 hence the number 124864 is divisible by 4. In this case remainder will be 1.

   4)124865 = 4)124864 r1 = 4)12/48/64 r1 =31216 r1.

?	**Identify remainder when following number divided by 4**		
	a. 9724 =	**b. 4296 =**	**c. 36577248 =**
	d. 102616 =	**e. 89676544 =**	**f. 6789384 =**
	g. 6789436 =	**h. 1672 =**	**i. 1568 =**
	j. 468 =	**k. 14946 =**	**l. 2346382 =**

4. Remainder when any number divided by 5 will be 1,2,3,4 or 0.

   E.g.: Any number 124868 will have remainder 3, when divided by 5 because last digit 8-5=3

?	**Identify remainder when following number divided by 5**		
	a. 9725 =	**b. 4294 =**	**c. 36577253 =**
	d. 102600 =	**e. 8967612 =**	**f. 6789301 =**
	g. 6789454 =	**h. 16720 =**	**i. 15595 =**
	j. 46854 =	**k. 149468 =**	**l. 23463802 =**

5. Remainder when any number divided by 6 will be 1,2,3,4,5 or 0.

   E.g.: Any number 124868 will have remainder 2, when divided by 6 because last digit 8 is even but digit sum = 2.

   (The number must be even and digit sum = 0, 3 or 6. Here remainder = digit sum of number – 0, 3 or 6 whichever nearest smaller number = 2-0 = 2)

?

Identify remainder when following numbers divided by 6

a. 9723 =	**b. 4296 =**	**c. 367245 =**
d. 102654 =	**e. 896769 =**	**f. 67932 =**
g. 6789453 =	**h. 1002 =**	**i. 1566 =**
j. 468 =	**k. 14949 =**	**l. 23412 =**

6. Remainder when any number divided by 7 will be 1,2,3,4,5,6 or 0. But exact remainder cannot be obtained unless divided.

7. Remainder when any number divided by 8 will be 1,2,3,4,5,6,7 or 0.

   E.g.: Any number 124868 will have remainder 4, when divided by 8. here 8x4+6x2+8=52 not divisible by 8 but 48 nearest to 52 is divisible by 8. So remainder is 52-48=4.

?	**Identify remainder when following number divided by 8**		
	a. 9714 =	**b. 4298 =**	**c. 36577242 =**
	d. 102619 =	**e. 8967604 =**	**f. 6789200 =**
	g. 6789434 =	**h. 1672 =**	**i. 1568 =**
	j. 468 =	**k. 14946 =**	**l. 2346382 =**

8. Remainder when any number divided by 9 will be 1,2,3,4,5,6,7,8 or 0.

   E.g.: Any number 124868 will have remainder 2, when divided by 9. here digit sum of 124868=2 , digit sum must be 0 for divisibility with 9 therefore remainder = 2-0=2.

?	**Identify remainder when following number divided by 9**		
	a. 9728 =	**b. 45397 =**	**c. 36577243 =**
	d. 102654 =	**e. 89676546 =**	**f. 6189343 =**
	g. 6789456 =	**h. 6007 =**	**i. 5464 =**
	j. 45261 =	**k. 14949 =**	**l. 2346381 =**

Short Method for Division

Division near base:

10, 100, 1000.....are called base. So division for those numbers close to bases can be through short method.

Numbers near 10

1. Suppose you need to divide a number 124868 by 9.

   Then we can use short method as follow.

   9 = 10 – **1**

9)	1	2	4	8	6	8	
1		1	3	7	6	8	
	1	**3**	**7**	**(15)**			
	1	3	8	6			
	1	**3**	**8**	**6**	**(12)**		
	1	**3**	**8**	**7**	**3**	**r(11)**	

= 13874 r2.

Another easy method for division by 9 is

Add subsequent digits with digits before and put in quotient.

9) 124868 then quotient

= 137(15)…

= 1386(12)..

= 13873(r11)

= 13874 r2

- put 1 =10-9 below 9 (row1)
- put first digit 1 of the number 124868 in quotient (row3)
- put 1 (put up in quotient) x1 (put up below 9) = 1 below 2 in the number (row 2).
- add 2 + 1 = 3 (row3)
- put 3 x 1(put up below 9) =3 below 4 in the number(row2).
- add 4 + 3 = 7 (row3)
- put 7x1(put up below 9)=7 below 8 in the number(row2).
- add 8 + 7=15 which is 9+6
- Transfer 1 in quotient to left(row4).
- Put 6x1=6 below 6 in the number (row2)
- add 6 + 6=12 which is 9+3
- Transfer 1 in quotient to left (row6)
- put 3x1=3 below 8 in the number (row2).
- add 8 + 3=11 which is 9+2 (row6)
- Transfer 1 in quotient to left digit so that remainder is 2.

?	**Divide following number by 9 through Short method**		
	a. 9728 =	**b. 45397 =**	**c. 36577243 =**
	d. 102654 =	**e. 89676546 =**	**f. 6189343 =**
	g. 6789456 =	**h. 6007 =**	**i. 5464 =**
	j. 45261 =	**k. 14949 =**	**l. 2346381 =**

2. Suppose you need to divide a number 124868 by 8 .

   Then we can use short method as follow.

   8 = 10 – 2

8) 1 2 4 8 6 8 2 2 8 8 0 12 **1 4 (12)** 1 5 4 (16) **1 5 6 0 6** **1 5 6 0 6 r(20)** = 15608 r4.	• put 2 =10-8 below 8 (row2) • put first digit 1 in quotient • put 1x2= 2 below 2 (row2) • add 2 + 2 = 4 (row3) • put 4 x 2 = 8 below 4 (row2) • add 8 + 4 = 12 which=8x1+4 • Transfer 1 in quotient to left • put 4x2=8 below 8 • add 8 + 8=16 which is=8x2+0 • Transfer 2 in quotient to left • put 0x2=0 below 6 • add 0 + 6=6 • now 6x2=12 below 8 (row2) add 8 + 12=20 which is=8x2+4

Transfer 2 to left digit in quotient so that remainder is 4.
Practice will reduce the calculation further.

?	**Divide following number by 8 through Short method**		
	a. 9714 =	**b. 4298 =**	**c. 36577242 =**
	d. 102619 =	**e. 8967604 =**	**f. 6789200 =**
	g. 6789434 =	**h. 1672 =**	**i. 1568 =**
	j. 468 =	**k. 14946 =**	**l 2346382 =**

3. Suppose you need to divide a number 124868 by 7.

   Then we can use short method as follow.

   7 = 10 – 3

7) 1 2 4 8 6 8 3 **3 15 15 6 15** **1 5 (19)** 1 7 5 (23) **1 7 8 2 (12)** **1 7 8 3 5 r(23)**	• put 3 =10-3 below 7 • put first digit 1 in quotient • now 1x3= 3 below 2 • add 2 + 3 = 5 • now 5 x 3 = 15 below 4 • add 4 + 15 = 19 which7+7+5 • Transfer 2 in quotient to left • now 5x3=15 below 8
• Transfer 3 in quotient to left • now 2x3=6 below 6 • add 6 + 6= 12 which is 7+5 • Transfer 1 in quotient to left • now 5x3=15 below 8 • add 8 + 15=23 which is 7x3+2 • Transfer 3 in quotient to left so that remainder is 2.	

?	**Divide following number by 7 through Short method**		
	a. 9714 =	**b. 4298 =**	**c. 3656 =**
	d. 1019 =	**e. 8604 =**	**f. 6678 =**
	g. 67864 =	**h. 1672 =**	**i. 1568 =**
	j. 468 =	**k. 14946 =**	**l. 2382 =**

4. Suppose you need to divide a number 124868 by 11. Then we can use short method as follow.

   11 = 10 + 1

11)	1	2	4	8	6	8
$\bar{1}$		$\bar{1}$	$\bar{1}$	$\bar{3}$	$\bar{5}$	$\bar{1}$
	1	**1**	**3**	**5**	**1**	**r7**

- put $\bar{1}$ below 11(row1)
- put first digit of the number 1 in quotient (row3)
- put 1x$\bar{1}$ = $\bar{1}$, below 2 (row2)
- add 2 + $\bar{1}$ = 1 (row3)
- put 1 x $\bar{1}$ = $\bar{1}$ below 4 (row2)
- add 4 + $\bar{1}$ = 3 (row2)
- put 3x$\bar{1}$ = $\bar{1}$ below 8 (row2)
- add 8 + $\bar{3}$ =5 (row3)
- put 5x$\bar{1}$ =$\bar{1}$ below 6 (row2)
- add 6 + $\bar{1}$ = 1 (row3)
- now 1x$\bar{1}$= $\bar{1}$ below 8 (row2)
- add 8 + $\bar{1}$ = 7(remainder in row3)

?	**Divide following number by 11 through Short method**		
	a. 276 =	**b. 6478 =**	**c. 39449 =**
	d. 76487 =	**e. 5228 =**	**f. 6799 =**
	g. 678945 =	**h. 1297 =**	**i. 1468 =**
	j. 4788 =	**k. 14947 =**	**l. 2346316 =**

5. Suppose you need to divide a number 124868 by 12.

   Then we can use short method as follow.

   12 = 10 + 2

$$\begin{array}{r\|rrrrrr} 12 & 1 & 2 & 4 & 8 & 6 & 8 \\ \bar{2} & & \bar{2} & \mathbf{0} & \bar{8} & \mathbf{0} & \overline{12} \\ \hline & \mathbf{1} & \mathbf{0} & \mathbf{4} & \mathbf{0} & \mathbf{6} & \mathbf{r}\bar{4} \end{array}$$ –4 cannot be remainder so take one from left digit to add 12 in –4 so that result is 10405r8	• put $\bar{2}$ below 12(row1) • put first digit of the number 1 in quotient (row3) • put 1x$\bar{2}$ = $\bar{2}$, below 2 (row2) • add 2 + $\bar{2}$ = 0 (row3) • put 0 x $\bar{2}$ =0 below 4 (row2) • add 4 + 0 = 4 (row2) • put 4x$\bar{2}$ = $\bar{8}$ below 8 (row2) • add 8 + $\bar{8}$ =0 (row3) • put 0x$\bar{2}$ = 0 below 6 (row2) • add 6 + 0 = 6 (row3)
• now 6x$\bar{2}$ = $\overline{12}$ below 8 (row2) • add 8 + $\overline{12}$ = $\bar{4}$ (remainder in row3)	

?	**Divide following number by 12 through Short method**		
	a. 276 =	**b. 6478 =**	**c. 39449 =**
	d. 76487 =	**e. 5228 =**	**f. 6799 =**
	g. 678945 =	**h. 1297 =**	**i. 1468 =**
	j. 4788 =	**k. 14947 =**	**l. 2346316 =**

6. Suppose you need to divide a number 124868 by 13.

   Then we can use short method as follow.

   12 = 10 + 3

<table>
<tr>
<td>
<table>
<tr><td>13)</td><td>1</td><td>2</td><td>4</td><td>8</td><td>6</td><td>8</td></tr>
<tr><td>$\overline{3}$</td><td></td><td>$\overline{3}$</td><td>3</td><td>$\overline{21}$</td><td>39</td><td>$\overline{18}$</td></tr>
<tr><td></td><td>1</td><td>$\overline{1}$</td><td>7</td><td>$\overline{13}$</td><td>(45)</td><td></td></tr>
<tr><td></td><td>1</td><td>1</td><td>7</td><td>$\overline{10}$</td><td>6</td><td>r$\overline{10}$</td></tr>
<tr><td></td><td></td><td>9</td><td>6</td><td>0</td><td>6</td><td>r$\overline{10}$</td></tr>
<tr><td></td><td></td><td>9</td><td>6</td><td>0</td><td>5</td><td>r3</td></tr>
</table>
–10 cannot be remainder therefore take one from 6 to add 13 to $\overline{7}$
=9605r3
</td>
<td>
• put $\overline{3}$ below 13(row1)
• put first digit of the number(1) in quotient (row3)
• put 1x$\overline{3}$ = $\overline{3}$ below 2 (row2)
• add 2 + $\overline{3}$ = $\overline{1}$(row3)
• put $\overline{1}$ x $\overline{3}$ = 3 below 4 (row2)
• add 4 + 3 = 7 (row3)
• put 7x$\overline{3}$ = $\overline{21}$ below 8(row2)
• add 8 + $\overline{21}$ = $\overline{13}$ (row3)
• put $\overline{13}$x$\overline{3}$ =39 below 6(row2)
• Here 45 = 13x3 +6
• Transfer 3 to left ($\overline{13}$ +3=$\overline{10}$ row4)
</td>
</tr>
<tr>
<td colspan="2">
• Put 6x$\overline{3}$ = $\overline{18}$ below 8 (row2)
• add 8 + $\overline{18}$ = $\overline{10}$ (row4)
• Now 1$\overline{1}$ = 9 and 7$\overline{10}$ = 60 (row 5)
</td>
</tr>
</table>

? **Divide following number by 13 through Short method**

a. 276 =	**b. 6478 =**	**c. 39449 =**
d. 76487 =	**e. 5228 =**	**f. 6799 =**
g. 678945 =	**h. 1297 =**	**i. 1468 =**
j. 4788 =	**k. 14947 =**	**l. 2346316 =**

Numbers near 100

1. Suppose you need to divide a number 124868 by 99

   Then we can use short method as follow.

   01 = 100 – 99

99) 12 48 68 01 12 60 12 60 r128 128 = 99 x 1 + 29 therefore transfer 1 to left digit and 29 as remainder **=1261r29**	• put 01 below 99 (row2) • put first two digit of the number (12) in quotient (row3) • put 12x01 = 12, below 48 (row2) • add 48 +12 = 60 (row3) • put 60 x 01 =600 below 68 (row2) • add 68 + 60 = 128 as remainder (row2)

?	**Divide following number by 99 through Short method**		
	a. 276 =	**b. 6478 =**	**c. 39449 =**
	d. 76487 =	**e. 5228 =**	**f. 6799 =**
	g. 678945 =	**h. 1297 =**	**i. 1468 =**
	j. 4788 =	**k. 14947 =**	**l. 2346316 =**

2. Suppose you need to divide a number 124868 by 98

   Then we can use short method as follow.

   02 = 100 – 98

98) 12 48 68 02 24 144 **12 72 r212** 212 = 98 x 2 + 16 therefore transfer 2 to left digit and 16 as remainder **=1274r16**	• put 02 below 98 (row2) • put first two digit of the number (12) in quotient (row3) • put 12x02 = 24, below 48 (row2) • add 48 +24 = 72 (row3) • put 72 x 02 =144 below 68 (row2) • add 68 + 144 = 212 as remainder (row2)

?	**Divide following number by 98 through Short method**		
	a. 276 =	**b. 6478 =**	**c. 39449 =**
	d. 76487=	**e. 5228 =**	**f. 6799 =**
	g. 678945 =	**h. 1297 =**	**i. 1468 =**
	j. 4788 =	**k. 14947 =**	**l. 2346316 =**

3. Suppose you need to divide a number 124868 by 101

Then we can use short method as follow.

101) 12 48 68 $\overline{01}$ $\overline{12}$ $\overline{36}$ 12 36 r32 **=1236r32**	• put $\overline{01}$ below 101 (row2) • put first two digit of the number (12) in quotient (row3) • put 12x$\overline{01}$ = $\overline{12}$, below 48 (row2) • add 48 +$\overline{12}$ = 36 (row3) • put 36 x $\overline{01}$ = $\overline{36}$ below 68 (row2) • add 68 + $\overline{36}$ = 32 as remainder (row2)

?	**Divide following number by 101 through Short method**		
	a. 276 =	**b. 6478 =**	**c. 39449 =**
	d. 76487 =	**e. 5228 =**	**f. 6799 =**
	g. 678945 =	**h. 1297 =**	**i. 1468 =**
	j. 4788 =	**k. 14947 =**	**l. 2346316 =**

4. Suppose you need to divide a number 124868 by 109

   Then we can use short method as follow.

$$\begin{array}{l\|lll} 101 & 12 & 48 & 68 \\ \overline{09} & & \overline{108} & 540 \\ \hline & 12 & \overline{60} & r608 \\ & 12 & \overline{55} & r63 \end{array}$$ Here 608 = 109 x 5 + 63, therefore add 6 to left and 2 as remainder. Now $12\overline{55}$ = **1145** **=1145r63**	• put $\overline{09}$ below 109 (row2) • put first two digit of the number (12) in quotient (row3) • put 12x$\overline{09}$ = $\overline{108}$, below 48 (row2) • add 48 + $\overline{108}$ = $\overline{60}$ (row3) • put $\overline{60}$ x $\overline{09}$ = $\overline{540}$ below 68 (row2) • add 68 + 540 = 608 as remainder (row2) • 608>109

?	**Divide following number by 109 through Short method**		
	a. 276 =	**b. 6478 =**	**c. 39449 =**
	d. 76487 =	**e. 5228 =**	**f. 6799 =**
	g. 678945 =	**h. 1297 =**	**i. 1468 =**
	j. 4788 =	**k. 14947 =**	**l. 2346316 =**

Divide any number by any other number with short method

1. Suppose you need to divide a number 124868 by 78

   Then we can use short method as follow.

78) 12 4 8 6 8 78) 124 468 006 8 1 6 00r 68 = 1600 r 68	• Here 124 = 78 x 1 + 46 • Put 1 in quotient and 46 as subscript of 8. So that number becomes 468. • 468 = 78 x 6 + 00 • Put 6 in quotient and 00 as subscript of 6. So that number becomes 0068 • 0068 = 78 x 00 + 68

2. Suppose you need to divide a number 124868 by 68

   Then we can use short method as follow.

68) 12 4 8 6 8 68) 124 568 246 428 1 8 3 6r20 = 1836 r 20	• Here 124 = 68 x 1 + 56 • Put 1 in quotient and 56 as subscript of 8. So that number becomes 568. • 568 = 68 x 8 + 24 • Put 8 in quotient and 24 as subscript of 6. So that number becomes 246 • 246 = 68 x 3 +42 • Put 3 in quotient and 42 as subscript of 8. So that number becomes 428. • 428 = 68 x 6 + 20 • Put 6 in quotient and 20 as remainder

CHAPTER 6

Fraction

A number of the form $\frac{p}{q}$ where $q \neq 0$ is called fraction. p=numerator, q=denominator.

If p<q then it is proper fraction.

E.g.: $\frac{1}{3}, \frac{2}{3}, \frac{5}{7}, \frac{8}{9}$

If p>q then it is improper fraction, which gets converted to mixed fraction by dividing numerator with denominator.

E.g.: $\frac{4}{3}, \frac{5}{3}, \frac{17}{7}, \frac{41}{9}$

A whole number and proper fraction together is called mixed fraction.

E.g.: Improper fraction = $\frac{4}{3}$, where $4 = 3 \times \mathbf{1} + 1$

Therefore, mixed fraction = $\mathbf{1}\frac{1}{3}$

Improper fraction = $\frac{41}{9}$, where 41= 9 x **4** + 5

Therefore, mixed fraction = $\mathbf{4}\frac{5}{9}$

Improper fraction = $\frac{17}{7}$, where 17 = 7 x 2 + 3

Therefore, mixed fraction = $2\frac{3}{7}$

Converting mixed fraction into improper fraction.

Mixed fraction = $\mathbf{4}\frac{5}{9}$ has whole number = 4 and proper fraction = $\frac{5}{9}$

Improper fraction

$$= \frac{\text{whole number} \times \text{denominator} + \text{numerator}}{\text{denominator}}$$

$$= \frac{4\times 9+5}{9} = \frac{36+5}{9} = \frac{41}{9}$$

Exercise: *Convert mixed fraction to improper fraction*

1. $4\frac{5}{9}$ **2.** $5\frac{7}{11}$ **3.** $11\frac{9}{11}$ **4.** $7\frac{7}{8}$

5. $12\frac{5}{18}$ **6.** $6\frac{3}{5}$ **7.** $21\frac{5}{6}$

Addition of Mixed Fraction

$5\frac{1}{3} + 6\frac{1}{2} = ?$

Short Method:

$5\frac{1}{3} = 5 + \frac{1}{3}$

$6\frac{1}{2} = 6 + \frac{1}{2}$

Therefore, we can write

$5\frac{1}{3} + 6\frac{1}{2} = 11 + (\frac{1}{3} + \frac{1}{2})$

Now take LCM of 3 and 2

$11 + (\frac{1}{3} + \frac{1}{2})$

$= 11 + \frac{1\times2+1\times3}{6}$

$= 11 + \frac{5}{6} = 11\frac{5}{6}$

School Method:

In school, we learn to solve such problem as

$5\frac{1}{3} = \frac{15}{3}$ (15= 5 x 3 + 1)

$6\frac{1}{2} = \frac{13}{2}$ (13 = 6 x 2 + 1)

Therefore, we write

$5\frac{1}{3} + 6\frac{1}{2} = \frac{15}{3} + \frac{13}{2}$

We take LCM of denominator (LCM of 3 and 2) which is 6.

$\frac{15}{3} + \frac{13}{2} = \frac{15\times2+13\times3}{6}$

$= \frac{69}{6} = 11\frac{3}{6}$

$5\frac{2}{3} + 6\frac{3}{2} = ?$

Ans: $\frac{13}{6}$

Short Method:

$5\frac{2}{3} + 6\frac{3}{2} = 11 + (\frac{2}{3} + \frac{3}{2})$

$= 11 + \frac{2\times2+3\times3}{6}$

$= 11 + \frac{13}{6} = 11 + 2\frac{1}{6} = 13\frac{1}{6}$

Here $\frac{13}{6} = 2 + \frac{1}{6}$

?	**Exercise:** *Add following fraction*
	1. $4\frac{5}{9}+3\frac{4}{3}$ **2.** $5\frac{7}{11}+6\frac{7}{11}$ **3.** $11\frac{9}{11}+5\frac{3}{4}$
	4. $7\frac{7}{8}+8\frac{9}{16}$ **5.** $12\frac{5}{18}+5\frac{7}{9}$ **6.** $6\frac{3}{5}+7\frac{3}{25}$
	7. $21\frac{5}{6}+8\frac{3}{9}$

Subtraction of Mixed Fraction

$5\frac{2}{3}-6\frac{3}{2}=?$	***Short Method:***
Ans: $-1\frac{5}{6}=-\frac{11}{6}$	$5\frac{2}{3}-6\frac{3}{2}=(5-6)+\left(\frac{2}{3}-\frac{3}{2}\right)$ $=(-1)+\left(\frac{2\times-3\times3}{6}\right)$ $=(-1)+\left(-\frac{5}{6}\right)=-1\frac{5}{6}$

?	**Exercise:** *Subtract following fraction*
	1. $4\frac{5}{9}-3\frac{4}{3}$ **2.** $5\frac{7}{11}-6\frac{7}{11}$ **3.** $11\frac{9}{11}-5\frac{3}{4}$
	4. $7\frac{7}{8}-8\frac{9}{16}$ **5.** $12\frac{5}{18}-5\frac{7}{9}$ **6.** $6\frac{3}{5}-7\frac{3}{25}$
	7. $21\frac{5}{6}-8\frac{3}{9}$

Comparing Fraction

Compare $\frac{3}{4}$ and $\frac{2}{3}$, which one is greater.

Here take L.C.M of denominator (4 and 3).

L.C.M = 12

Now multiply 12 in both fraction.

$\frac{3}{4} \times 12 = 9$ and $\frac{2}{3} \times 12 = 8$

Here 9>8

So $\frac{3}{4} > \frac{2}{3}$

Compare $\frac{3}{4}, \frac{2}{3}$ and $\frac{5}{7}$ and write in ascending order

Here take L.C.M of numbers in denominator (4, 3 and 7).

L.C.M = 84

Now multiply 84 in all the fraction.

$\frac{3}{4} \times 84 = 63, \frac{2}{3} \times 84 = 56$ and $\frac{5}{7} \times 84 = 60$

Here 63 > 60 > 56

So $\frac{3}{4} > \frac{5}{7} > \frac{2}{3}$

?

Exercise: *Compare following fraction*

1. $\frac{5}{9}, \frac{3}{4}$
2. $\frac{7}{9}, \frac{5}{6}$
3. $\frac{9}{11}, \frac{13}{22}$
4. $\frac{7}{8}, \frac{13}{16}, \frac{5}{4}$
5. $\frac{5}{7}, \frac{11}{14}, \frac{19}{21}$
6. $\frac{3}{5}, \frac{3}{25}$
7. $\frac{5}{6}, \frac{7}{9}$

Multiplication of Fraction

$$\frac{3}{4}\times\frac{2}{3}=?$$

Here numerator multiply with numerator and denominator multiply with denominator. Numerator and denominator divides each other.

3 divides 3 and 2 divides 4

$$\frac{3}{4}\times\frac{2}{3}=\frac{1}{2}\times\frac{1}{1}=\frac{1}{2}$$

?

Exercise: *Multiply following fraction*

1. $\frac{5}{9}\times\frac{3}{4}$ **2.** $\frac{7}{9}\times\frac{5}{6}$ **3.** $\frac{9}{11}\times\frac{13}{22}$

4. $\frac{7}{8}\times\frac{13}{16}$ **5.** $\frac{5}{7}\times\frac{11}{14}$ **6.** $\frac{3}{5}\times\frac{3}{25}$

7. $\frac{5}{6}\times\frac{7}{9}$

Division of Fraction

$$\frac{3}{4}\div\frac{2}{3}=?$$

Here to find division, multiply $\frac{3}{4}$ with reciprocal of $\frac{2}{3}$ i.e $\frac{3}{2}$

$$\frac{3}{4}\div\frac{2}{3}=\frac{3}{4}\times\frac{3}{2}=\frac{9}{8}=1\frac{1}{8}$$

?

Exercise: *Multiply following fraction*

1. $\frac{5}{9} \div \frac{3}{4}$ **2.** $\frac{7}{9} \div \frac{5}{6}$ **3.** $\frac{9}{11} \div \frac{13}{22}$

4. $\frac{7}{8} \div \frac{13}{16}$ **5.** $\frac{5}{7} \div \frac{11}{14}$ **6.** $\frac{3}{5} \div \frac{3}{25}$

7. $\frac{5}{6} \div \frac{7}{9}$

Fraction into Percentage

Fractions can be converted in percentage (% = 1/100).

Convert $\frac{4}{25}$ into %.

Here fraction is proper (denominator > numerator), multiply both numerator and denominator by 100 so that denominator in result is 100.

Now $\frac{4}{25} = \frac{4 \times 100}{25 \times 100} = \frac{8}{100} = 8\%$

Convert $\frac{108}{25}$ into %.

Here fraction is improper, denominator < numerator.

$108 = 25 \times 4 + 8 = 400\%$ of $25 + 8$

Now $\frac{108}{25} = \frac{400\% \text{ of } 25 + 8}{25} = \frac{400\% \text{ of } 25}{25} + \frac{8}{25}$

$= 400\ \% + \frac{8 \times 100}{25 \times 100} = 400\ \% + \frac{32}{100}$

$= 400\% + 32\% = 432\%$

Note: 100% of 25 = 25, 200% of 25 = 50, 300% of 25 = 75.....

?	**Exercise:** *Convert following fraction into parentage* **1.** $\frac{3}{4}$ **2.** $\frac{8}{25}$ **3.** $\frac{17}{5}$ **4.** $\frac{17}{20}$

A number = 100 % of the number 2 = 100% of 2 5 =100% of 5 30 = 100% of 30 etc.....	10 % of any number = number/10 10% of 30=30/10= 3 10% of 50 = 50/10= 5 etc.

Similarly,

1 % of any number = number/100

1% of 30 = 30/100 = 0.3

1% of 50 = 50/100 = 0.5 etc.

Convert $\frac{99}{30}$ into %.

Here $\frac{99}{30} = \frac{30\times3+9}{30} = \frac{3\times30+3+3+3}{30}$

$= 3 \times 100\,\% + 10\,\% + 10\% + 10\%$

$= 300\% + 30\% = 330\%$

Convert $\frac{87}{30}$ into %.

$\frac{87}{30} = \frac{2\times30+3x(3+3+3)}{30}$

$= 2 \times 100\% + 3 \times (10\% + 10\% + 10\%)$

$= 200\% + 3 \times (30\%) = 200\% + 90\% = 290\%$

Convert $\frac{46.8}{9}$ into %.

$$\frac{48}{9} = \frac{5\times9+1.8}{9} = \frac{5\times9+0.9+0.9}{9}$$

$$= 500\% + 10\% + 10\% = 520\%$$

(Since 10% of 9 = 0.9 and 100% of 9 = 9)

Convert $\frac{49}{15}$ into %.

$$\frac{51}{15} = \frac{3\times15+6}{15} = 300\% + \frac{6}{15} = 300\% + \frac{6}{15}$$

$$= 300\% + \frac{4\times1.5}{15} = 300\% + 4 \times 10\%$$

$$= 300\% + 40\% = 340\%$$

?

Exercise: *Convert following fraction into parentage*

1. $\frac{3}{4}$ 2. $\frac{8}{25}$ 3. $\frac{17}{5}$ 4. $\frac{25}{20}$

CHAPTER 7
Decimals

Terminating Decimal or Recurring Decimal

Decimals can be Terminating Decimal or Recurring Decimal.

Terminating Decimal: Terminating decimal is a decimal number that has digits that do not go on forever. i.e does not repeat for ever.

For example:

0.15 (it has two decimal digits)

7.2371 (it has four decimal digits)

Recurring Decimal: In contrast to terminating decimals a Recurring Decimal has digits that reoccur forever.

For example:

1/3 = 0.333... (the 3 repeats forever) is a Recurring Decimal, not a Terminating Decimal.

Identifying a fraction which will be terminating or recurring.

A fraction of the form $\frac{p}{q}$ $(q \neq 0)$ will be terminating when q (denominator) can be expressed in the form of prime factor of 2 and 5 only.

If $q = 2^n \times 5^m$ where n and m are any whole number then decimal will be terminating.

If $q \neq 2^n \times 5^m$ then decimal will be recurring.

E.g.: $\frac{9}{8} = 1.125$ which is terminating.

Here $8 = 2^3 \times 5^0$

$\frac{14}{15} = 0.933333.....$ non-terminating or recurring decimal

Here $15 = 3 \times 5$ not in the form of $2^n \times 5^m$, here factor is 3 other than 2 and 5 so the decimal will be non-terminating or recurring.

Exercise: *Without division, identify the fractions whether terminating or recurring.*

1. $\frac{3}{4}$ **2.** $\frac{8}{25}$ **3.** $\frac{17}{5}$

4. $\frac{21}{20}$ **5.** $\frac{14}{15}$ **6.** $\frac{20}{21}$

7. $\frac{36}{125}$

Made in United States
North Haven, CT
06 November 2026

21866983R00065